Studies in Computational Intelligence

Volume 1210

The series “Studies in Computational Intelligence” (SCI) publishes new developments and advances in the various areas of computational intelligence—quickly and with a high quality. The intent is to cover the theory, applications, and design methods of computational intelligence, as embedded in the fields of engineering, computer science, physics and life sciences, as well as the methodologies behind them. The series contains monographs, lecture notes and edited volumes in computational intelligence spanning the areas of neural networks, connectionist systems, genetic algorithms, evolutionary computation, artificial intelligence, cellular automata, self-organizing systems, soft computing, fuzzy systems, and hybrid intelligent systems. Of particular value to both the contributors and the readership are the short publication timeframe and the world-wide distribution, which enable both wide and rapid dissemination of research output.

Indexed by SCOPUS, DBLP, WTI AG (Switzerland), zbMATH, SCImago.

All books published in the series are submitted for consideration in Web of Science.

Erik Cuevas · Nahum Aguirre ·
Oscar Barba-Toscano · Mario Vásquez-Franco

Advanced Metaheuristics: Novel Approaches for Complex Problem Solving

Springer

Erik Cuevas
CUCEI, Universidad de Guadalajara
Guadalajara, Jalisco, Mexico

Nahum Aguirre
CUCEI, Universidad de Guadalajara
Guadalajara, Mexico

Oscar Barba-Toscano
CUCEI, Universidad de Guadalajara
Guadalajara, Mexico

Mario Vásquez-Franco
CUCEI, Universidad de Guadalajara
Guadalajara, Mexico

ISSN 1860-949X ISSN 1860-9503 (electronic)
Studies in Computational Intelligence
ISBN 978-3-031-89283-7 ISBN 978-3-031-89284-4 (eBook)
https://doi.org/10.1007/978-3-031-89284-4

This Springer imprint is published by the registered company Springer Nature Switzerland AG
The registered company address is: Gewerbestrasse 11, 6330 Cham, Switzerland

Preface

Optimization is the process of identifying the most effective solution or result from a range of available alternatives, typically to enhance or reduce a particular objective or standard. This concept is essential across numerous disciplines such as mathematics, engineering, economics, and computer science.

Metaheuristic approaches are robust problem-solving strategies extensively employed in optimization to tackle intricate, multidimensional, or computationally demanding issues. These techniques offer a broad framework for investigating and enhancing potential solutions to discover near optimal or optimal outcomes within vast solution spaces. In contrast to conventional optimization algorithms, metaheuristics exhibit greater flexibility and adaptability. They derive inspiration from natural processes and phenomena, such as evolution, simulated annealing, or swarm behavior, to steer the search for solutions. Through iterative refinement and the exploration of solutions while avoiding local optima, metaheuristics can efficiently address a diverse array of optimization challenges across various fields. This makes them invaluable tools for solving real-world problems in which traditional methods may prove impractical or inadequate.

No single metaheuristic approach can obtain the best results for solving all optimization problems across the board. Consequently, the creation of new algorithms is crucial. By developing innovative methods and mechanisms, we can investigate new strategies that deviate from conventional approaches. These novel techniques enable us to address the specific intricacies and unique features of particular problems, thereby providing customized solutions that enhance both efficiency and effectiveness.

The scientific field has recently seen the introduction of numerous novel metaheuristic algorithms, which have shown promise owing to their impressive results in various optimization scenarios. However, it is crucial to recognize that many of these algorithms are still in their early stages of development. Although their effectiveness has been successfully demonstrated using mathematical functions, the true test of their capabilities lies in their application to complex, real-world challenges. Only when these algorithms are confronted with the intricacies and difficulties presented by practical, multidimensional optimization tasks will their full potential and limitations become evident. Thorough testing and validation of such intricate problems are crucial to assess their feasibility and effectiveness in addressing the diverse and complex optimization challenges encountered across different fields.

Balancing exploration and exploitation is a key challenge in metaheuristic algorithms for effective optimization. Exploration is vital for finding new and potentially superior solutions, whereas exploitation focuses on refining and utilizing known promising solutions. Striking the right balance is difficult because of the complex trade-offs involved. Excessive emphasis on exploration can result in slow convergence and excessive computational overhead, whereas excessive focus on exploitation may lead to premature convergence to suboptimal solutions. Finding the optimal balance is problem-specific and often requires a deep understanding of the characteristics of the problem. The dynamic nature of many real-world problems adds another layer of complexity. As a result, researchers have used heuristic strategies and adaptive mechanisms to address this dilemma, as it remains a key challenge in optimizing complex and multifaceted problems.

In recent years, the field of optimization has witnessed the introduction of numerous metaheuristic algorithms, each purporting to offer novel approaches for solving complex problems. However, many of these algorithms are, in essence, variations of the existing ones, often differing only in the metaphor or inspiration they employ while recycling previously used operators with minimal modifications. This practice leads to several issues: it engenders a false sense of innovation, diminishes scientific rigor in the field, and complicates the selection process for practitioners confronted with a plethora of seemingly distinct algorithms that are, in reality, functionally similar. Moreover, this trend impedes meaningful progress by diverting attention away from the development of novel techniques and superficial changes that offer little or no improvement in performance.

This book presents and explains new metaheuristic algorithms with interesting search mechanisms that distinguish them from traditional methods. These innovative approaches offer new strategies for addressing complex optimization problems. By examining these novel techniques, this book underscores the benefits of moving beyond conventional algorithms. New operators often capture various problem characteristics and explore potential solutions more thoroughly, potentially yielding better results in complex scenarios. Studying and considering these alternative methods enables researchers and practitioners to overcome the constraints of standard algorithms, thereby creating opportunities to solve previously challenging or inefficient problems. Furthermore, these innovative approaches can stimulate new ideas and contribute to the development of more sophisticated techniques, driving significant advancements in optimization research.

The target readership of this book encompasses a wide range of individuals with diverse expertise and interests, although certain groups will find it especially pertinent and useful. The primary audience consists of researchers and scholars in the fields of optimization, artificial intelligence, and metaheuristic algorithms. These readers will value the comprehensive analysis of cutting-edge metaheuristic algorithms and a thorough assessment of established techniques, as this information can further their research and contribute to the creation of groundbreaking optimization methods. Moreover, industry professionals working in sectors that rely heavily on optimization, such as logistics, engineering, finance, and data science, will find this book extremely valuable. It provides practical insights into how optimization algorithms perform on real-world challenges, helping them to choose the most appropriate approaches for their specific applications. Lastly, the book serves as an excellent resource for graduate students and advanced undergraduates studying optimization and algorithm design. By offering both theoretical foundations and practical examples, they enhanced their comprehension of metaheuristic methods and their application to complex optimization problems.

The book explores a set of strategies designed to enhance metaheuristic algorithms, focusing on key aspects such as initialization methods, the integration of Evolutionary Game Theory for developing new search mechanisms, and the application of learning concepts to refine evolutionary operators. Additionally, it highlights the importance of diversity and opposition in preventing premature convergence and improving algorithmic efficiency. These strategies collectively contribute to the development of more adaptive and robust optimization techniques. To provide a structured exploration of these concepts, the book is organized into six chapters, each offering a detailed discussion of the methodologies, theoretical foundations, and practical implementations that drive advancements in metaheuristic optimization.

Chapter 1 introduces the fundamental concepts of learning strategies within metaheuristic algorithms, offering a detailed review of their key characteristics and advantages. It examines how these strategies enhance the search process by enabling algorithms to adapt and improve over time. A particular focus is given to the role of both individual and collective behavior in shaping the efficiency of metaheuristic techniques. By analyzing how agents interact, share information, and influence one another, the chapter highlights their contribution to exploring and exploiting solution spaces more effectively. This foundational discussion sets the stage for understanding how learning mechanisms drive optimization performance.

Chapter 1 serves as an introduction to the world of metaheuristic methods, laying the foundation for the rest of the book by discussing their core structures and key features. This opening chapter is designed to familiarize readers with the fundamental principles that underpin metaheuristic algorithms, including their ability to navigate and optimize within complex search spaces.

Chapter 2 provides a comprehensive overview of the various types of metaheuristic algorithms found in the literature, offering a detailed analysis of their underlying mechanisms and the specific operators that define their functionality. It examines the general structure that governs these algorithms, highlighting the key components that drive their optimization processes. Additionally, the chapter explores different search strategies employed in diverse optimization environments, focusing on the most influential and widely used metaheuristic techniques. By understanding how these algorithms operate and adapt, readers gain valuable insights into their effectiveness and applications in solving complex optimization problems.

Chapter 3 introduces a novel approach to population initialization in metaheuristic algorithms using the Gibbs sampling methodology. This technique generates samples for each decision variable of a particle based on a conditional distribution, considering previously obtained samples. By iteratively producing diverse samples, the methodology reduces correlations between individuals in the population, leading to a more uniform spatial distribution. This improved distribution enhances the algorithm's exploration capabilities, particularly in high-dimensional search spaces, ultimately contributing to better optimization performance and solution quality.

Chapter 4 introduces a metaheuristic algorithm that dynamically adjusts its search strategy throughout the optimization process, enhancing efficiency and adaptability. Built upon the principles of Evolutionary Strategies (ES), the algorithm leverages their self-adaptive properties to refine its performance. Additionally, it integrates concepts from Evolutionary Game Theory (EGT), enabling a more sophisticated strategy adaptation process by considering both individual agent behavior and interactions within the population. Unlike traditional EGT-based approaches, this method expands adaptation mechanisms by incorporating insights from top-performing individuals alongside pairwise evaluations. By leveraging this enhanced learning framework, the algorithm achieves faster convergence toward optimal strategies, improving overall optimization effectiveness.

Chapter 5 explores the fundamental concepts of learning strategies in metaheuristic algorithms, providing an in-depth analysis of their key characteristics and advantages. Traditional metaheuristic methods often emphasize the best solution found collectively while overlooking individual agent strategies, which can lead to premature convergence. To address this limitation, the chapter introduces a novel approach that balances individual and social learning in search agents. Each agent operates using two complementary strategies: an independent exploration mechanism and a socially driven search based on the best-known collective solution. This framework treats the optimization process as a learning problem, where agents dynamically adjust the balance between these strategies. A randomized counter assigned to each agent regulates the frequency of strategy selection, ensuring diverse search patterns and fostering adaptability. By integrating these mechanisms, the proposed method enhances search efficiency and improves performance in solving complex optimization problems.

Finally, Chap. 6 presents a hybridized version of the Cheetah Optimizer, incorporating Opposition-Based Learning (OBL) and a diversity control mechanism to enhance its performance. By integrating dynamic strategy selection with diversity regulation, the algorithm significantly improves its exploration and exploitation capabilities. This hybrid approach allows the optimizer to maintain a well-balanced search process, accelerating convergence while reducing the risk of becoming trapped in local optima. The combination of OBL and diversity control ensures a broader and more effective search space exploration, leading to more efficient optimization results.

We wish to thank many people who were involved in the writing of this book. We express our gratitude to Dr. Thomas Ditzinger who supported this book project.

Guadalajara, Mexico
February 2025

Erik Cuevas
Nahum Aguirre
Oscar Barba-Toscano
Mario Vásquez-Franco

Competing Interests The authors have no competing interests to declare that are relevant to the content of this manuscript.

Contents

Chapter 1
Optimization

1.1 Introduction

Throughout life, people face situations that require them to make decisions between several possible options. People acquire homes, vehicles, clothing, food or other material goods; they can choose a profession, a partner for work or a life partner, and maybe a different city to live in. In all these decisions, people tend to prefer better options over worse ones. However, we are often aware of our limitations and constraints, so we don't always select the absolute best option in every aspect. Instead, we make decisions that we consider the optimal result, considering our priorities, resources, and constraints. For example, while a large, beautiful, and well-located house might seem the best answer, we may sacrifice the size of the house for a better location near the city or vice versa based on our circumstances. In essence, we don't always choose the "best" option but rather the "optimal" one—what is best within the given conditions, like priorities, resources and constraints. However, choosing an alternative that meets all the conditions can be complicated in certain cases. This process can be broadly referred to as Optimization.

Optimization, a branch of mathematics, focuses on this decision-making process [1]. Another way to define optimization is the process of deciding what cost the individual is willing to incur or what sacrifices the individual is prepared to make to achieve specific benefits. In mathematics, a similar concept applies: every outcome typically involves a cost, represented by a variable that reflects the trade-off required to achieve a desired result. This cost is not necessarily negative; rather, it is simply the effort, resource, or compromise necessary to obtain the expected outcome. This discipline has gained relevance by facilitating the search for solutions that satisfy the greatest number of possible requirements. Therefore, optimization is present in multiple aspects of our lives [2], from organizing personal agendas and improving economic systems and game strategies to optimizing a country's health system.

E. Cuevas et al., *Advanced Metaheuristics: Novel Approaches for Complex Problem Solving*, Studies in Computational Intelligence 1210,
https://doi.org/10.1007/978-3-031-89284-4_1

Consider the example of optimizing the construction of a bridge. This process involves systematically enhancing various aspects of design, materials, and construction methods to achieve the best results within specified constraints. Engineers might optimize the structural design to reduce material usage while maximizing load-bearing capacity and ensuring safety. This includes selecting the ideal combination of materials, such as steel and concrete, to balance durability and cost-effectiveness. Furthermore, construction schedules can be optimized to minimize time and labor costs, promoting efficiency. By employing optimization techniques like finite element analysis and metaheuristic algorithms, engineers can evaluate numerous design variables and constraints, ultimately creating a bridge that is more durable, economical, and sustainable.

Another example is that in the automotive market, car designers provide options like sports cars, which are built to offer high acceleration capabilities. These vehicles typically feature powerful engines and are designed to be lightweight, which explains why sports cars are usually small but equipped with robust engines. To maximize acceleration, designers focus on maximizing engine power while minimizing the car's weight. However, sports car owners generally do not prioritize low fuel consumption or economic maintenance. In contrast, vehicles designed for low fuel consumption are also small but often feature low-power engines to achieve this goal. Large family cars, on the other hand, are built to balance acceleration and fuel efficiency while accommodating several passengers, sometimes with three rows of seats. These vehicles are significantly larger and equipped with engines that are not optimized for high acceleration but rather for meeting the practical demands of family transportation. Cargo trucks, while equipped with very powerful engines to handle heavy loads, tend to have the lowest acceleration due to their large size and weight.

However, these examples do not strictly mean that optimization is only present in these areas because, as mentioned above, optimization is present in individuals or consumers. In other words, some people prefer not to invest many resources when deciding; for example, they prefer it to be functional regardless of the design it may have or vice versa; they can choose to save a little on other actions to invest more in other purchases. In all these possible scenarios, there is a common factor: the individual tries to optimize their resources (economic, time, etc.) in order to obtain a good. It is thanks to this that optimization can be considered to be present in everything, including people's daily lives. Due to this, optimization is a widely studied branch [3].

In other words, physical situations can be represented in mathematical terms, which are usually called "mathematical models" [4]. Once they are modeled, they can be optimized using different techniques we will describe in this book.

From a mathematical perspective, optimization seeks to identify the "best solution" to a problem, considering a series of restrictions within a search space defined by an objective function. The objective function is a mathematical representation of a variable that one wishes to optimize and that depends on other input variables, limitations or conditions under which the problem must be solved. Generally, this type of problem is represented by the following equation, which allows determining the function's minimum or maximum, depending on the user's preference.

$$\begin{matrix} maximize\, f(x) & \\ x \in \boldsymbol{X} \end{matrix} \; or \; \begin{matrix} minimize\, f(x) \\ x \in \boldsymbol{X} \end{matrix} \tag{1.1}$$

In Eq. 1.1, $f(x)$ is the objective function that needs to be either maximized or minimized. It represents the performance measure to be optimized and is defined in terms of one or more variables; x is the vector of decisive variables that represents the choices to be made in the problem. The goal is to determine the values of these variables that optimize the objective function; and $\boldsymbol{X}$ is the search space or solution space that represents the set of all potential solutions that satisfy the constraints. Therefore, an optimization problem can be formulated as a process where we want to find the optimal value x that minimizes or maximizes the objective function. It is due to this property that optimization has gained fame by providing solutions to different challenges in scientific fields.

In the field of optimization, a wide variety of concepts are necessary to efficiently explore the process of solving an optimization problem. Some of these concepts are the limits in the search space, local optima and global optima. From this, the search space is limited by the lower l_i and upper u_i limits of each of the decision variables such that it can be represented as in the next equation:

$$\boldsymbol{X} = \left\{ x \in R^d \,|\, l_i \leq x_i \leq u_i, i = 1, \ldots, d \right\} \tag{1.2}$$

where d represents the number of decision variables.

In most optimization problems, the solutions must satisfy certain restrictions, which can be expressed as either inequalities or equalities. Mathematically speaking, these constraints can be represented as:

$$\begin{aligned} g_i(x) \leq 0 \quad i = 1, \ldots, m \\ h_j(x) = 0 \quad j = 1, \ldots, p \end{aligned} \tag{1.3}$$

Finally, the solution x^* is considered optimal if:

$$\begin{aligned} f(x^*) \leq f(x) \forall x \in X \; in\, case\, of\, minimization \\ f(x^*) \geq f(x) \forall x \in X \; in\, case\, of\, maximization \end{aligned} \tag{1.4}$$

Another way to illustrate these concepts is through the metaphor of a mountainous region [5]. In this scenario, each peak represents a local optimum since there are no higher points in its vicinity. However, only the highest peak of the mountain is considered the global optimum. In many optimization problems, the real challenge lies in finding that global optimum due to the existence of multiple local optima in the search space.

- *Optimization*
  - *Objective*
    - *One Objective*
    - *Multi − Objective*
  - *Restriction*
    - *Without Restiction*
    - *Restricted*
  - *Type*
    - *Unimodal*
    - *Multimodal*
  - *Form of the Function*
    - *Linear*
    - *Non − Linear*
      - *Quadratic*
      - ...
  - *Variables*
    - *Discrete*
      - *Integra*
      - ...
    - *Continue*
    - *Mixed*
  - *Determination*
    - *Deterministic*
    - *Stochastics*

Fig. 1.1 Classification of optimization problems

Optimization problems can have different approaches depending on the exercises to be solved since each of these exercises can be similar or completely different, from problems with restrictions to discrete variable problems. A summary of this diversity can be observed in the following figure:

In Fig. 1.1, various optimization approaches are presented depending on the characteristics of the problem. Below, we briefly discuss some of the types of problems:

- Based on the existence of Restrictions
  Generally speaking, if the set of restrictions $X = \mathbb{R}^d$, the optimization problem is called an unconstrained optimization problem and can be defined as follows:

$$\min_{x \in \mathbb{R}^d} f(x) \tag{1.5}$$

  while a constrained optimization problem can be defined as follows:

$$\begin{gathered} \min_{x \in \mathbb{R}^d} f(x) \\ \text{such that } c_i(x) = 0,\ i \in E \\ c_i(x) \geq 0,\ i \in I \end{gathered} \tag{1.6}$$

  where I and E represent, respectively, the set of inequality constraints and equality constraints [6]. Based on the above we can define $c_i(x)$ and $(i = 1, \ldots, m \in E \cup I)$ as restriction functions.
- Based on the Design Variables
  As seen in Fig. 1.2, we can divide the optimization problems based on the design variables into continuous optimization problems and discrete optimization problems. Continuous optimization problems are based on finding the optimal value

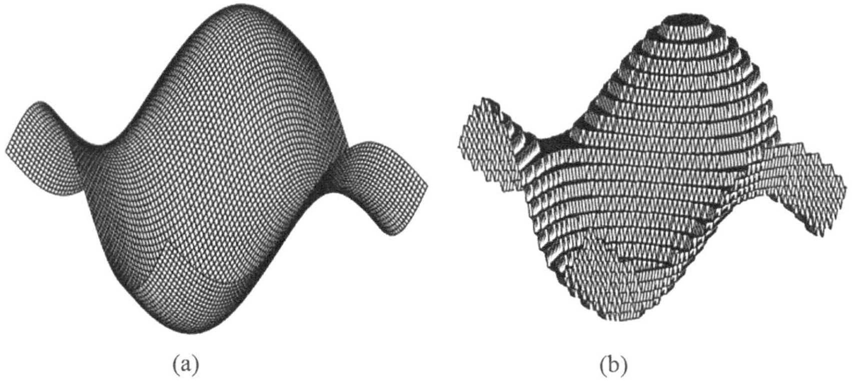

Fig. 1.2 Difference between a continuous three-dimensional sinusoidal function (**a**) versus the same function discretized (**b**)

of a continuous function and can be multimodal or constrained problems. On the other hand, discrete optimization problems are restricted to using only sets of discrete values, such as integers [7].

- Based on the structure of the objective function
  Objective functions guide the optimization process by defining a search space through mathematical models, so we can classify these based on their mathematical structure as follows:

  Linear Functions: This is defined as an objective function in which all its variables and restrictions are of first-order [8], which is defined in univariate mode mathematically as follows:

$$f(x) = mx + b \tag{1.7}$$

  Quadratic Functions: This is defined as an objective function in which its highest term is degree 2, and all its restrictions are of first order, defined mathematically in univariate mode as follows:

$$f(x) = ax^2 + bx + c \tag{1.8}$$

  Non-Linear Functions: These objective functions are defined when one or more of their terms and constraints involve non-linear terms such as exponentials, square roots, logarithms, trigonometric functions, etc. [9].

  Convex Functions: This term refers to the functions that are defined on a convex set C, and that, for any two points x and y members of C, and for each t in [0, 1], the following condition is fulfilled [10]:

$$f(tx + (1-t)y) \leq tf(x) + (1-t)f(y) \tag{1.9}$$

Concave Functions: They are defined as those functions that are defined on a convex set C, and that for any points x and y in the domain C, and for any t in [0, 1], the following condition is fulfilled [11]:

$$f(tx + (1 - t)y) \geq tf(x) + (1 - t)f(y) \quad (1.10)$$

- Based on the number of Optima in the objective Function
  Optimization problems can be classified based on the number of optima that can be found in an objective function into two categories:
  Unimodal Functions: These are functions with a single global minimum or maximum point [12], being defined mathematically as a function in which there exists a point x_m such that:

$$f(x)\ is\ decreasing\ in\ (-\infty, x_m]\ and\ growing\ in\ [x_m, \infty)\ for\ minimization$$
$$f(x)\ is\ growing\ in\ (-\infty, x_m]\ and\ decresing\ in\ [x_m, \infty)\ for\ maximization$$

  Multimodal Functions: These are functions in which there is a local minimum or maximum, as seen in Fig. 1.3, having multiple points where the function changes direction [13], expressed mathematically as a function defined in an interval I in which there are at least two points x_1 and x_2 such that:

$$There\ is\ a\ gap\ (a, b) \subset I,\ where\ f(x)\ is\ growing\ for\ x \in (a, x_1),$$

$$decresing\ for\ x \in (x_1, x_2),\ and\ growing\ again\ for\ x \in (x_2, b).$$

- Based on the Number of objective functions
  Within this category, we can concentrate on problems that have one or more objective functions.

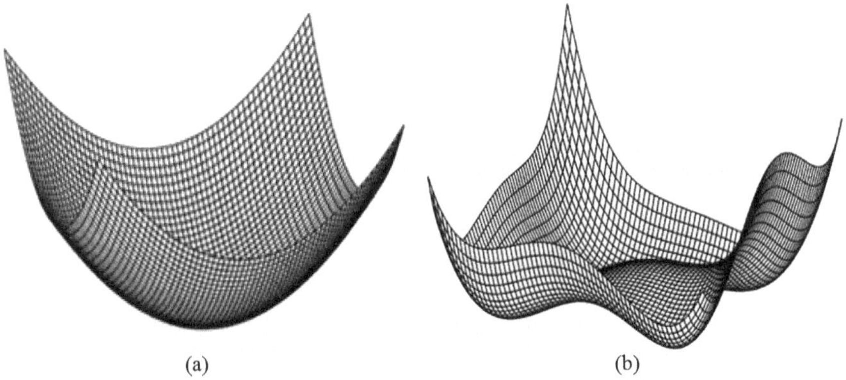

Fig. 1.3 Difference between an unimodal function (**a**) in contrast to a multimodal function (**b**)

$$\text{Optimization}\begin{cases} \text{Deterministic}\begin{cases} \text{Linear Programming}\\ \text{Non}-\text{linear Programmin}\\ \dots\\ \text{Based on the Gradient}\\ \text{Gradient Free}\end{cases}\\ \text{Stochastics}\begin{cases} \text{Heuristic}\\ \text{Metaheuristic}\begin{cases}\text{Based on Population}\\ \dots\\ \text{Based on Trayectory}\end{cases}\end{cases}\end{cases}$$

Fig. 1.4 Algorithms deterministic and stochastics

One Objective: This classification includes problems in which it is only necessary to optimize a single objective function.

Multi-Objective: This classification includes optimization problems that simultaneously deal with several objective functions to find a solution that adequately satisfies them using Pareto efficiency as a means [14].

- Based on the Deterministic nature of the Variables
  Within this category, we can concentrate on the optimization problems in two different ways:

  Stochastic Nature: This refers to experiments whose behavior is determined by probability, where there is no certainty of the result obtained [15].

  Deterministic Nature: This refers to systems in which all the variables are known, allowing the response of the system to be predicted accurately [16].

However, in the optimization area, these last two main methods stand out for their use in optimization algorithms for solving these problems, as shown in Fig. 1.4: the deterministic methods and the stochastic methods [17].

Optimization algorithms are processes created to find the optimal point in a given objective function, which we can classify according to behavior as:

- **Deterministic Algorithms**: These are algorithms whose output is completely conditioned by the input data they receive; that is, if we always enter the same data, then we will obtain the same result in each execution that is carried out. Therefore, we can see that there is no randomness factor in this type of algorithm. These algorithms are quite reliable and have a wide variety of use cases due to their low computational resource consumption and consistency in the results; however, this is conditioned on their search space meeting strict parameters [18].
- **Stochastic Algorithms**: These algorithms incorporate some randomness mechanism in their search processes. They are far from deterministic algorithms since even if we introduce the same input data, we will always obtain a different output. Therefore, any process of this nature will involve probabilistic mechanisms to analyze its results [19].

In the literature, these methods are also known as classical methods and metaheuristic methods. Classical methods use mathematical tools to achieve greater accuracy in solutions. However, as the complexity of the problems increases, these methods may become impractical for generating answers. In contrast, metaheuristic methods overcome this limitation by focusing on finding approximate solutions. Although they do not seek absolute accuracy, they offer results close to the optimum with significantly shorter processing times.

1.2 Classical Methods

The objective function may be non-linear with respect to the decision variables, which adds a level of complexity to the problem. To address this difficulty, optimization methods employ iterative processes designed to explore the search space efficiently. Classical methods use the function's gradient to generate new candidate solutions. Due to the operation characteristics of classical optimization algorithms, they require that the objective function meets two essential requirements: it must be twice differentiable and unimodal [20]. Due to these characteristics with which classical methods work, they also have several advantages, including being able to reduce an n-dimensional problem to a one-dimensional problem; another advantage is that due to the greater amount of information that these types of problems have, it is possible to achieve convergence at a higher speed. Over the years, several authors developed some algorithms based on the derivative of the objective function. Some of these algorithms are briefly described below:

- Gradient Descent Method
  This method, also known as the gradient method, is one of the first techniques used for the minimization of multidimensional objective functions [21]. This method forms the basis for several other optimization algorithms because, despite its slow convergence, it is the most used for optimizing non-linear functions.

  In this method, starting from an initial point x_0, the decision vector is modified iteratively until after a certain number of iterations *Niter*, the optimal solution x^* can be tentatively found. The following Eq. 1.11 determines such iterative modification:

  $$x(k+1) = x(k) - \alpha g(f(x)) \tag{1.11}$$

  where k represents the current iteration and α represents the search step size. Furthermore, the term $g(f(x))$ represents the gradient of the objective function $f(x)$. The gradient g of a function $f(x)$ at point x expresses the direction in which the maximum growth is found. In the case of minimization, it is the opposite direction.
- Newtonian Method
  In this method, the downward direction $\alpha g(f(x))$ can be determined by using the second derivative of the objective function $f(x)$. Thus, if the function is analyzed considering the position x_0 as the starting point, the objective function can be approximated as a quadratic function. Because it has a quadratic form

with respect to x, it is possible to find the minimum point by the traditional method of differentiating the equation. The result is the following linear Eq. 1.12:

$$g(f(x)) + H(f(x))(x - x_0) = 0 \tag{1.12}$$

where $H(f(x))$ represents the Hessian matrix consisting of the second partial derivative of the function. This value can also be obtained iteratively as shown in the following Eq. 1.13:

$$x(k+1) = x(k) - H^{-1}g \tag{1.13}$$

where the magnitude $\|H^{-1}g\|$ is known as Newton's step. Likewise, its direction represents Newton's direction. From Eqs. 1.11 to 1.13 the following representation can be established:

$$H_k(x_{k+1} - x_k) = g_{k+1} - g_k \tag{1.14}$$

Although these classical methods have proven useful for solving unimodal and doubly differentiable problems, they have several disadvantages because they depend on the derivative, and there are problems where this can be computationally expensive, thus resulting in a problem that these methods cannot solve.

1.2.1 Example of an Optimization Problem

A farmer wants to build a fence for his cattle. He has a 100-m fence and wants to build a rectangular corral adjacent to a wall (it should be added that this wall does not need a fence). The problem to be solved is to maximize the area of the corral with the available length. To better understand the problem, Fig. 1.5 represents the fence, with X and Y representing the dimensions.

In this exercise, we have two variables: x, which represents the side parallel to the wall, and y, which represents the two sides perpendicular to the wall. A restriction is the total length of the fence is 100 m, which gives us the following equation:

$$x + 2y = 100 \tag{1.15}$$

Solving for y we get the following:

$$y = \frac{100 - x}{2} \tag{1.16}$$

Since the area of the corral is defined as:

$$A(x) = x * y \tag{1.17}$$

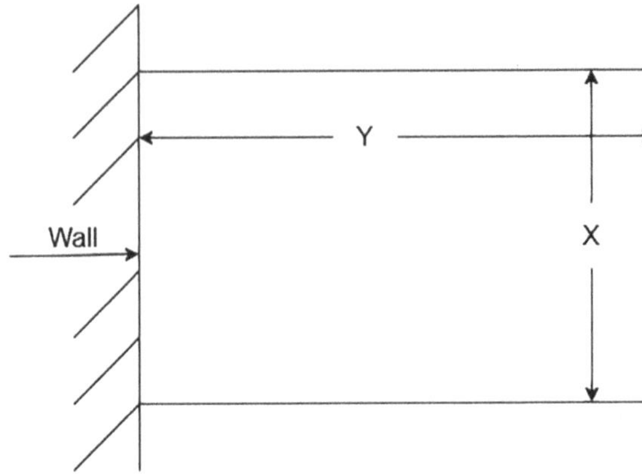

Fig. 1.5 Representation of the area to be fenced

By replacing y we obtain the objective function to be optimized (area of the corral) which is determined by the following:

$$A(x) = x * \frac{100 - x}{2} = 50x - \frac{x^2}{2} \tag{1.18}$$

With all these equations, the optimization problem can be described, i.e., the objective function in its canonical form, which initially expresses whether maximization or minimization is necessary, followed by "subject to" and the restrictions that the problem has. The canonical form can be expressed as follows:

$$max f(x) = 50x - \frac{x^2}{2} \tag{1.19}$$

Subject to:

$$x + 2y = 100 \tag{1.20}$$

$$x > 0 \tag{1.21}$$

$$y > 0 \tag{1.22}$$

In the case of the farmer, the dimensions of the corral may be two numbers, x and y; however, by using equality in the first constraint, it was possible to express y in terms of x to define the objective function only in terms of one variable. The constraints (1.21) and (1.22) may seem obvious; however, in an optimization formulation, it's important to include all possible information because sometimes the problem modeling can be done by someone expert, another person specializing in optimization or by software. Code 1.1 solves the described problem using the gradient method:

```
%Code 1.1 - maximize area, gradient method
%Objective function (area) and its derivative
A = @(x) 50*x - (x.^2)/2;        % Area of the fence
dA = @(x) 50 - x;                % Derivative of the area function

%Initial Parameters
x = 10;                 %Starting point
alpha = 0.1;            %Learning rate
max_iter = 100;         %Maximun number of iterations

%Descending gradient
for i = 1:max_iter
    grad = dA(x);               %Calculate the gradient
    x_new = x + alpha * grad; %Update x

    %Limit x to the valid range [0, 100]
    if x_new > 100
        x_new = 100;
    elseif x_new < 0
        x_new = 0;
    end

    x = x_new; %Update x for next iteration
end

% Results
y = (100 - x) / 2;
fprintf('Optimal Dimensions: x = %.6f, y = %.6f\n', x, y);
fprintf('Max Area: A(x) = %.6f\n', A(x));
```

When running this code, the algorithm maximizes the area of the corral by adjusting the value of x at each iteration. At the optimal point, it is noticeable that the solution is $x = 50$ and $y = 25$. Thus, a maximum area of 1250 m will be achieved while respecting the restrictions assigned by the problem.

On the other hand, the Newtonian method can solve the same problem while demonstrating other characteristics of this method; the differences between both methods will be discussed later. This method finds an extreme using the following formula:

$$x_{n+1} = x_n - \frac{f'(x)}{f''(x)} \tag{1.23}$$

Therefore, to solve the same problem of the farmer's fence, we can use the same objective function obtained previously (Eq. 1.16). In contrast, Eq. 1.23 tells us that we need its first and second derivative for this process, which can be observed in the following equations:

$$A'(x) = 50 - x \tag{1.24}$$

$$A''(x) = -1 \tag{1.25}$$

Now that we have the necessary information to solve this exercise, code 1.2 solves the problem described using the Newtonian method:

```
%Code 1.2 - maximize area, newtonian method
%Objective function (area) and its derivative
A = @(x) 50*x - (x.^2)/2;    %Area of the fence
dA = @(x) 50 - x;            %First derivative
ddA = @(x) -1;               %Second derivative

%Initial Parameters
x = 10;              %Initial Point
max_iter = 100;      %Maximun number of iterations

%Newtonian Method
for i = 1:max_iter
    grad = dA(x);                  %Calculate the gradient
    hessian = ddA(x);              %Calculate the second derivative
    x_new = x - grad / hessian; %Update x using Newtonian Method

    x = x_new; %Update x for next iteration
end

% Results
y = (100 - x) / 2;
fprintf('Optimal Dimensions: x = %.6f, y = %.6f\n', x, y);
fprintf('Max Area: A(x) = %.6f\n', A(x));
```

Due to the objective function's parabolic nature, the Newtonian method converges faster to the optimal response than the gradient method. This is also because, from the first iteration, the Newtonian method uses both the slope of the first derivative and the curvature of the second derivative to move towards the maximum. In other words, the Newtonian method has two essential advantages compared to the gradient method:

- **Speed**: Newton's method converges faster because it uses additional information (the second derivative).
- **Adaptability**: Newton's method adjusts the step according to the curvature of the function, avoiding problems with poorly chosen learning rates.

In conclusion, classical methods are easy to use when the problem to be optimized allows it, since with the correct information, it is possible to solve it without much effort. However, the exercise used for this example is considered a simple and basic exercise because not all optimization problems meet the conditions demonstrated in this exercise, especially the fact of being doubly derivable. It is because of this that metaheuristic methods are used to solve optimization problems, and therefore, they are mostly mentioned in the literature.

1.3 Metaheuristic Methods

Metaheuristic methods [22] are considered derivative-free procedures because they do not require the objective function to be twice derivable or unimodal. These approaches incorporate global optimization techniques capable of addressing various types of optimization challenges, including non-convex, nonlinear, and multimodal problems, while accounting for both linear and nonlinear constraints as well as continuous or discrete decision variables. Because of this, metaheuristic methods, such as global optimization algorithms, can understand nonconvex, nonlinear, and multimodal problems, which are subject to linear or nonlinear constraints with discrete or continuous variables.

Metaheuristic techniques have a diverse history due to the evolution of various parameters, such as technology or the nature of optimization problems. As industrial processes become increasingly complex, the demand for innovative optimization methods continues to grow. This is notable for metaheuristic techniques, which stand out for their ability to operate without prior knowledge or any assumption about the optimization problem. Unlike classical methods based on gradients of objective functions, metaheuristic approaches are especially suited to handle highly nonlinear systems or objective functions often encountered in engineering. These functions frequently involve noise and lack explicit deterministic expressions, requiring experimental or simulation-based evaluations for potential solutions. In the literature, there are exercises called "engineering problems" [23], which simulate these challenges to test the capabilities of the new algorithms that emerge in the literature. All these new challenges have established metaheuristic methods as powerful alternatives in the field of optimization.

A metaheuristic technique is a general method for solving optimization problems. It uses the objective function efficiently and abstractly, typically without going deeply into its mathematical properties. Unlike classical methods, metaheuristic methods do not require further information about the objective function since they are used as if they were "black boxes" [24]. They solve the problem in a "black box" fashion, which denotes their flexibility and popularity.

Metaheuristic methods take advantage of the information obtained from all solutions evaluated during the optimization process to infer the structure of the optimization problem. This information is then used to generate new potential solutions, with the expectation that these will outperform previously found solutions; in other words, this is being done to improve the quality of future solutions. Over the years, many metaheuristic techniques have shown remarkably promising results due to the possible modifications that can be made to already existing methods or to the various creative techniques that are born to produce a completely new algorithm. These methods are often inspired by models found in biological, natural or social systems, which, in a certain sense, can be interpreted as optimization problems in themselves.

Within the metaheuristic methods, there are various classifications, with the possibility that they work and/or are based on a population or trajectory, and within these, there are other classifications. Thus, population algorithms are those algorithms that use a search with several initial points from each individual in the population. With this, they evaluate the objective function in search of candidate solutions that can lead to the global optimum [25]. Bio-inspired algorithms, also known as those inspired by nature, are those based on the behavior and evolution of animals and nature itself [26]. Trajectory-based algorithms are those that work based on a single solution at any time and cover only local searches [27]. In addition, swarm-based algorithms are those that use simple particles or agents that interact with each other and with the environment locally [28]. Despite this, there is no unified vision of the classifications, which leads to the same classification being used for several algorithms. An example of this occurs with the term swarm because sometimes it is used for population algorithms that have a group behavior model [29]. Figure 1.6 summarizes the best-known algorithms for some of the classifications mentioned above. The following references include various studies on the algorithms' growth and evolution [30–33].

For example, inspired by the cooperative behaviors observed in natural systems, various optimization methods have been developed to address complex problems. For example, the Artificial Bee Colony (ABC) [34] algorithm draws inspiration from the foraging behavior of honeybees, where individual bees work together to explore and exploit food sources, ultimately identifying optimal solutions. Similarly, the Particle Swarm Optimization (PSO) [35] algorithm is inspired by the social behavior of bird flocks or fish schools. In this method, particles representing potential solutions navigate the search space by updating their positions based on their own experiences

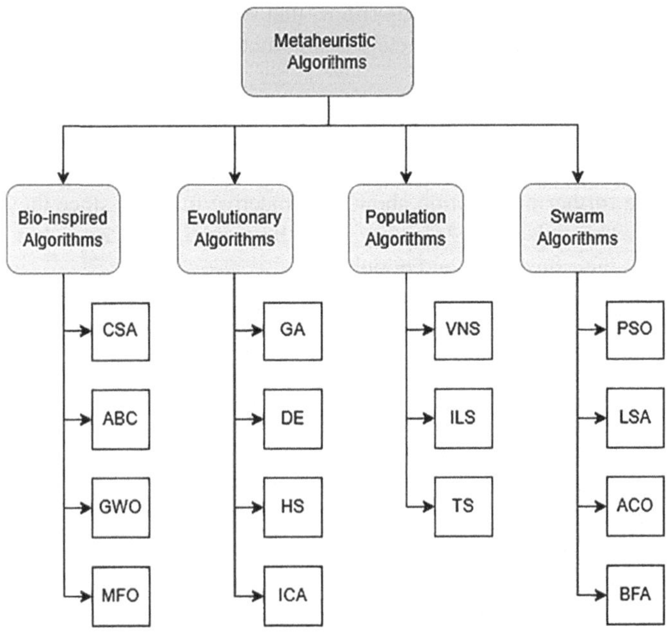

Fig. 1.6 Classification of metaheuristic algorithms

and the experiences of neighboring particles, facilitating the discovery of optimal solutions. Metaheuristic techniques encompass a broad spectrum of strategies. The Differential Evolution (DE) [36] algorithm, for example, is derived from a framework rooted in vector algebra combined with principles of evolutionary computation. Similarly, Genetic Algorithms (GA) [37] are inspired by the genetic mechanisms of biological species, mimicking processes such as selection, crossover, and mutation to solve optimization problems.

However, the number of different metaheuristic algorithms in the literature has increased over the years because, as mentioned in the "no free lunch" theorem [38], no heuristic or metaheuristic solves all optimization problems. After all, each algorithm has its own characteristics that favor the solution of a specific type of problem.

1.3.1 Common Structure of a Metaheuristic Method

A metaheuristic approach is generally defined as a search strategy that seeks to emulate behaviors observed in biological, natural, or social systems. These methods are designed to efficiently explore and exploit the search space, often drawing inspiration from processes such as evolution, animal behaviors, or social interactions. A typical metaheuristic algorithm comprises several key components:

- **Solution representation**: This is the first key component, since it is from here that all the information about the problem to be optimized is taken and we begin to work with that information. Therefore, it is necessary that the entire structure of the solution is established in order to work properly. This information may vary from problem to problem; however, it is necessary that the algorithm can adapt to these changes, from the number of dimensions that the problem has to the size of the population with which it is going to work, in order to be a competitive algorithm in the literature.
- **Initialization**: This component is considered crucial because it establishes an initial population or set of candidate solutions with which the algorithm will be working. This is usually carried out with a degree of randomness in several of the metaheuristic algorithms; that is, some methods allow the initial population to be created from random points within the search space. However, since it is considered a crucial stage for the functioning of the algorithm, it is possible to improve the algorithm's performance by improving this stage. Therefore, initialization methods seek to use the information of the search space so that the initial population is created in promising areas, thus achieving a better exploration of the search space and avoiding the complications that randomness can bring. This helps to avoid stagnation zones and improves the convergence of the algorithm.
- **Evaluation**: This component is possibly the core of a metaheuristic algorithm, since it is here where the objective function is evaluated with the values established by the population and it is determined if the action performed is the correct one for optimization, that is, the objective function is evaluated and it is analyzed if

the result of the new candidate solution is better compared to its predecessor. This stage is linked to two crucial components of a metaheuristic algorithm, mutation and selection, since by mutating the results obtained from the initial population and its consequences, new possible solutions are obtained that are then evaluated in the objective function, and if the result obtained is better than its predecessor then it is selected to form part of the new population which will serve as a guide for the next iteration.

- **Exploration and exploitation mechanisms**: The metaheuristic approach is designed to progressively improve the quality of the candidate solutions throughout the optimization process, also known as exploration and exploitation. Exploitation involves refining the solutions identified during exploration to enhance their quality. While exploration focuses on searching the entire solution space for new possibilities, exploitation hones in on the most promising solutions uncovered during exploration, making local adjustments to optimize them further. Achieving an effective balance between exploration and exploitation is essential, as it allows for identifying promising regions of the solution space and then fine-tuning the best results for optimal outcomes.
- **Termination criteria**: The termination criterion, also known as stopping criterion, is a key component because a metaheuristic algorithm, being a computational process established in a cycle, needs to establish where to stop it, otherwise it can continue with its work until it consumes many computational resources and could generate a problem. This is why in the different algorithms, not only in the metaheuristic ones, but different stopping criteria are also established, where the algorithm reaches an optimal solution for the problem to be solved. This stopping criterion can often be established based on factors such as a fixed number of iterations, computational restrictions or convergence criteria. It is possible to establish this criterion depending on the problem to be performed, an example of this could be image processing where by using a comparison metric the criterion can be established in terms of iterations, or if a comparison criterion is not available, it can be established in a non-referenced metric, which establishes the stopping criterion in a value established by the author where it is estimated that the result is optimal for the problem.

These components work in harmony, enabling metaheuristic algorithms to tackle complex optimization problems effectively, often producing good, if not always optimal, solutions. A metaheuristic approach operates by maintaining a diverse population of candidate solutions that evolves continuously throughout the optimization process. This population is refined iteratively, with new solutions replacing existing ones to explore the solution space more thoroughly and increase the likelihood of finding superior outcomes. Each solution is evaluated for quality using a cost function, which informs the selection process by determining which candidates persist and contribute to future generations. To navigate the solution space efficiently, various operational mechanisms are employed, enabling the intelligent generation of high-quality solutions. Combined, these components create a robust and adaptable framework for tackling complex optimization challenges.

Over time, it is expected to become more efficient in solving the problem. During execution, the algorithm's defined functions generate new solutions whose quality is refined with each iteration. In metaheuristic methods, the quality of a solution is directly tied to its effectiveness in addressing the optimization problem, driving the population toward the global optimal solution. This robust mechanism has enabled metaheuristic techniques to tackle complex engineering problems across various disciplines. Most metaheuristic methods [39] address the challenge of discovering global solutions for nonlinear constrained optimization problems, often within box constraints.

References

1. D. Simon, *Evolutionary Optimization Algorithms*. (Wiley, 2013)
2. A.D. Belegundu, T.R. Chandrupatla, *Optimization Concepts and Applications in Engineering* (Cambridge University Press, 2019)
3. S.S. Rao, *Engineering Optimization: Theory and Practice* (Wiley, 2019)
4. R. Sioshansi, A.J. Conejo, *Optimization in Engineering*, vol. 120 (Springer International Publishing, Cham, 2017)
5. R. Wehrens, L.M.C. Buydens, Classical and nonclassical optimization methods," *Encyclopedia of Analytical Chemistry* (2000), pp. 9678–9689
6. W. Sun, Y.-X. Yuan, *Optimization Theory and Methods: Nonlinear Programming*, vol. 1 (Springer Science & Business Media, 2006)
7. R.G. Parker, R.L. Rardin, *Discrete Optimization* (Elsevier, 2014)
8. D. Bertsimas, J.N. Tsitsiklis, *Introduction to Linear Optimization*, vol. 6 (Athena Scientific Belmont, MA, 1997)
9. A. Ruszczynski, *Nonlinear Optimization* (Princeton University Press, 2011)
10. D. Bertsekas, *Convex Optimization Theory*, vol. 1 (Athena Scientific, 2009)
11. D. Bertsekas, A. Nedic, A. Ozdaglar, *Convex Analysis and Optimization*, vol. 1 (Athena Scientific, 2003)
12. I.A. Ibragimov, On the composition of unimodal distributions. Theory Probab. Appl. **1**(2), 255–260 (1956)
13. M. Preuss, S. Wessing, Measuring multimodal optimization solution sets with a view to multi-objective techniques, in *EVOLVE-A Bridge between Probability, Set Oriented Numerics, and Evolutionary Computation IV: International Conference held at Leiden University*, 10–13 July 2013 (Springer, 2013), pp. 123–137
14. M.T.M. Emmerich, A.H. Deutz, A tutorial on multiobjective optimization: fundamentals and evolutionary methods. Nat. Comput. **17**, 585–609 (2018)
15. J.C. Spall, *Introduction to Stochastic Search and Optimization: Estimation, Simulation, and Control* (Wiley, 2005)
16. M. Cavazzuti, M. Cavazzuti, Deterministic optimization, in *Optimization methods: From Theory to Design Scientific and Technological Aspects in Mechanics* (2013), pp. 77–102
17. I. Zelinka, V. Snasael, A. Abraham, *Handbook of Optimization: From Classical to Modern Approach*, vol. 38 (Springer Science & Business Media, 2012)
18. L. Breiman, A. Cutler, A deterministic algorithm for global optimization. Math. Program. **58**(1), 179–199 (1993)
19. A. Shapiro, Y. Wardi, Convergence analysis of stochastic algorithms. Math. Oper. Res. **21**(3), 615–628 (1996)
20. P. Venkataraman, *Applied Optimization with MATLAB Programming* (Wiley, 2009)
21. J.E. Dennis, A brief introduction to quasi-Newton methods. Numer. Anal. **22**, 19–52 (1978)

22. B.-B. Michael, *Nonlinear Optimization with Engineering Applications* (Springer, Berlin/ Heidelberg, Germany, 2008)
23. X.-S. Yang, *Engineering Optimization: An Introduction with Metaheuristic Applications* (Wiley, 2010)
24. C. Blum, A. Roli, Metaheuristics in combinatorial optimization: overview and conceptual comparison. ACM Comput. Surv. (CSUR) **35**(3), 268–308 (2003)
25. Z. Beheshti, S. Mariyam, H. Shamsuddin, A review of population-based meta-heuristic algorithm (2013), www.i-csrs.org
26. S. Binitha, S.S. Sathya, A survey of bio inspired optimization algorithms. Int. J. Soft Comput. Eng. **2**(2), 137–151 (2012)
27. P. Hansen, N. Mladenović, J. Brimberg, J.A.M. Pérez, *Variable Neighborhood Search* (Springer, 2019)
28. E. Bonabeau, M. Dorigo, G. Theraulaz, *Swarm Intelligence: From Natural to Artificial Systems* (Oxford University Press, 1999)
29. E.V.C. Jiménez, D.A.O. Navarro, M.A.D. Cortés, J.V.O. Enciso, *Optimización: Algoritmos programados con MATLAB* (Alpha Editorial, 2016)
30. P. Agrawal, H.F. Abutarboush, T. Ganesh, A.W. Mohamed, Metaheuristic algorithms on feature selection: a survey of one decade of research (2009–2019). IEEE Access **9**, 26766–26791 (2021). https://doi.org/10.1109/ACCESS.2021.3056407
31. H. Malik, Atif, I. Puneet, J. Sanjay, A. Farhad, I. Bakhsh, Studies in computational intelligence 916 metaheuristic and evolutionary computation: algorithms and applications (2021), http://www.springer.com/series/7092
32. M. Abdel-Basset, L. Abdel-Fatah, A.K. Sangaiah, Metaheuristic algorithms: a comprehensive review, in *Computational Intelligence for Multimedia Big Data on the Cloud with Engineering Applications* (2018), pp. 185–231
33. V. Tomar, M. Bansal, P. Singh, Metaheuristic algorithms for optimization: a brief review. MDPI AG, March 2024 (2024), p. 238. https://doi.org/10.3390/engproc2023059238
34. D. Karaboga, B. Basturk, A powerful and efficient algorithm for numerical function optimization: artificial bee colony (ABC) algorithm. J. Glob. Optim. **39**, 459–471 (2007)
35. J. Kennedy, R. Eberhart, Particle swarm optimization, in *Proceedings of ICNN'95-International Conference on Neural Networks* (IEEE, 1995), pp. 1942–1948
36. R. Storn, K. Price, *Differential Evolution-A Simple and Efficient Heuristic for Global Optimization over Continuous Spaces* (Kluwer Academic Publishers, 1997)
37. J.H. Holland, Genetic algorithms and adaptation, in *Adaptive Control of Ill-Defined Systems* (1984), pp. 317–333
38. Y.-C. Ho, D.L. Pepyne, Simple explanation of the no free lunch theorem of optimization, in *Proceedings of the 40th IEEE Conference on Decision and Control (Cat. No. 01CH37228)* (IEEE, 2001), pp. 4409–4414
39. E. Cuevas, F. Wario, V. Osuna-Enciso, D. Zaldivar, M. Pérez-Cisneros, Fast algorithm for multiple-circle detection on images using learning automata. IET Image Process **6**(8), 1124–1135 (2012)

Chapter 2
Metaheuristic Algorithms

2.1 Introduction

Metaheuristic algorithms, as the prefix "meta" indicates, have a higher level of abstraction than heuristic algorithms, as they are algorithms designed with a generic structure that covers a large number of optimization problems [1]. This class of algorithms retains its characteristics, and we find a solution with a good quality-cost ratio by adjusting parameters depending on the characteristics of the problem we are dealing with [2].

These algorithms can be classified by different characteristics, such as their source of inspiration, population size, particle motion, and number of parameters, among others. However, the most referenced classification is by their source of inspiration.

Within the classification of metaheuristic algorithms by their source of inspiration, we can find different subcategories, as shown in the following illustration [3] (Fig. 2.1).

2.2 Evolutionary Algorithms

Within this classification of metaheuristic algorithms, we can find that they are mainly inspired by the postulates of biological evolution, which have robust methods with the ability to prevail over time thanks to their updating and adaptation techniques [4].

The two main operators on which these types of algorithms are based are mutation, crossover and selection. Mutation is a process of individual change where randomness plays a fundamental role. Crossover consists of combining the information of two or more individuals to generate offspring with shared characteristics that help to have greater diversity in the population. Selection is based on an elitist principle where

E. Cuevas et al., *Advanced Metaheuristics: Novel Approaches for Complex Problem Solving*, Studies in Computational Intelligence 1210,
https://doi.org/10.1007/978-3-031-89284-4_2

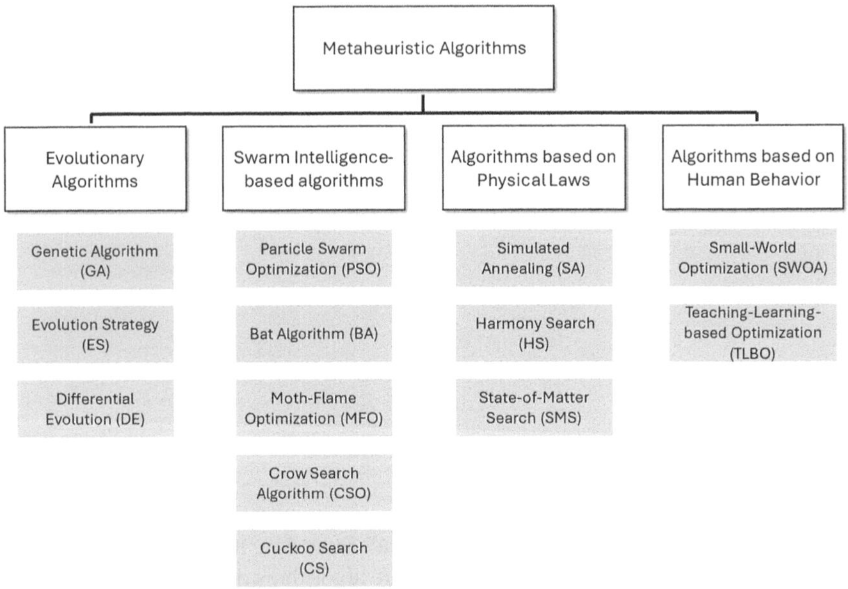

Fig. 2.1 Classification of metaheuristic algorithms according to their source of inspiration

only the best individuals will prevail over time, while individuals with worse results will tend to disappear [5].

2.2.1 Genetic Algorithm

Proposed by John Henry Holland in the 1960s, this model is based on the principles of evolution and gene reproduction, which were later adapted to solve optimization problems as adaptive methods. This algorithm encodes a solution, which we call a chromosome, through the application of the crossover and mutation operators [6]. The crossover operator consists of the recombination of two or more parent chromosomes to obtain a daughter chromosome resulting from the combination of genetic information. Mutation consists of modifying the daughter chromosomes so that they can develop their characteristics from a genetic base.

Due to the nature of its crossover and selection operator, it is necessary to have a population of chromosomes so that there is a diversity of solutions, as well as a selection principle based on natural selection, where based on the quality of solution in the objective function, will be its probability of surviving and transmitting its information to the next generation [7].

The computational implementation of this algorithm consists of 4 steps shown below:

1. **Initialization**: Like most metaheuristic algorithms, the generation of individuals in a search space is necessary to start evaluating the objective function and verifying potential zones. This process is performed under the following scheme:

$$X = lb + R \cdot (ub - lb) \tag{2.1}$$

   where X represents each individual, ub and lb represent the upper and lower bounds of the search space, respectively, and R is a uniformly distributed random number.
2. **Selection**: Individuals are chosen according to their aptitudes so that the best ones prevail and the weakest ones disappear. This process can be carried out in different ways. However, the most common is to make a selection based on a classification of the value of each individual in the objective function.
3. **Crossover**: Combinations are made based on the selected individuals to create offspring individuals with genetic characteristics of the parent individuals.
4. **Mutation**: The individual, with a certain probability, will modify its attributes to develop characteristics of its own and have a higher probability of survival.

Step 1 is performed only at the start of the algorithm, while steps 2–4 will be executed iteratively until a stopping criterion is reached.

2.2.2 *Evolutionary Strategies*

These models are based on the biological processes of adaptation that occur in nature, where individuals mutate to adapt to the environment and improve their characteristics. The individual resulting from the mutation will be called "offspring," and it will replace the "parent" individual only if the offspring obtained better characteristics due to the mutation. If the offspring individual does not perform better, then it will be discarded, and the parent individual will mutate again [8].

The first method developed from this family of algorithms is known as ES(1 + 1), where "ES" refers to "Evolutionary Strategy" and "1 + 1" refers to the type of mutation to be performed, where a parent individual in the population that will mutate into a child individual is taken into account [9]. Thanks to the results obtained from the simplicity of this mechanism, improved versions were developed, modifying the number of parent individuals, the number of offspring individuals that can result from mutation, as well as the implementation of dynamic strategies so that mutations are more likely to improve the qualities of individuals.

The computational implementation of this algorithm consists of the four steps shown below:

1. **Initialization:** An individual is generated at a point in the search space as follows:

$$X_p = lb + R \cdot (ub - lb) \tag{2.2}$$

where X_p represents the generated individual, *ub* and *lb* represent the upper and lower bounds, respectively, and R is a uniformly distributed random number. The variance parameter σ^2 is also initialized.

After this process, the evaluation of the individual in the objective function is performed, where:

$$fX_p = f(X_p) \tag{2.3}$$

2. **Generation of mutation vector**: A mutation vector is generated as shown below:

$$\rho = N(0, \sigma^2) \tag{2.4}$$

where ρ is the mutation vector, and N is a normally distributed random number with mean *zero* and standard deviation σ^2.

3. **Mutation**: The individual performs a process of adaptation of its attributes, modifying them as follows:

$$X_h = X_p + \rho \tag{2.5}$$

where X_h is the offspring resulting from the mutation.

After this process, the evaluation process of the individual child is performed in the objective function, where:

$$fX_h = f(X_h) \tag{2.6}$$

4. **Selection**: A comparison is made, where:

$$X_p = \begin{cases} X_p, fX_h \geq fX_p \\ X_h, fX_h < fX_p \end{cases} \tag{2.7}$$

Steps 1 and 2 are performed only at the start of the algorithm. In contrast, steps 3 and 4 are performed iteratively until a stopping criterion is reached, where a limited number of iterations is usually set.

2.2.3 *Differential Evolution*

Proposed by Price [10], the Differential Evolution (DE) algorithm is an evolutionary optimization method that proposes the calculation and generation of vectors based on the addition of the weighted difference between two members of a population. This algorithm simulates the natural evolution of a population to find an optimal solution by performing evolutionary operations such as crossover and mutation.

This algorithm bases its operation on a population with evolutionary operators, which have key parameters that regulate the crossover rate and the scale factor. The optimization process of the differential evolution algorithm can be implemented in a series of 4 steps detailed below:

1. **Initialization**: A population is created within the search space as follows:

$$x_i = lb + R \cdot (ub - lb) \tag{2.8}$$

where x_i represents the ith generated individual from a population of N number of particles at iteration k, ub and lb represent the upper and lower bounds, respectively, and R is a uniformly distributed random number.
2. **Mutation**: Once a population has been created within the search space, the mutation operator is in charge of creating a new solution vector based on the information contained in three selected individuals, called a mutant vector. There are several variants of the mutation operator; however, the original idea of mutant vector generation can be visualized below:

$$v = x_{r3} + F \cdot (x_{r1} - x_{r2}) \tag{2.9}$$

where x_{r1}, x_{r2} y x_{r3} (with $r1, r2, r3 \in \{1,2, \ldots, N\}i$ and $r1 \neq r2 \neq r3 \neq i$) are vectors of randomly selected parameters, and F is a scaling factor that controls the magnitude of the vector resulting from the difference between x_{r1} and x_{r2} with a range from *zero* to 2. It is important to have a correct handling of the scaling factor F so that the algorithm does not have a premature convergence nor is unable to find the global optimum but that there is a correct balance to achieve the best results based on the dimensions of the search space (Fig. 2.2).

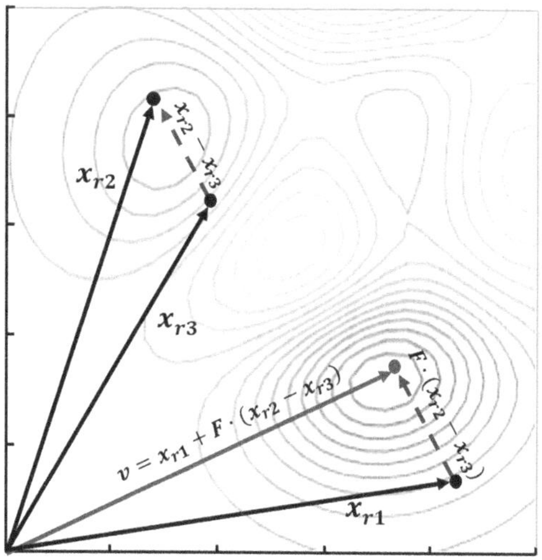

Fig. 2.2 Generation of mutant vector based on selected vectors

3. **Crossover**: While the mutation operator focuses on generating new solution vectors from information in other vectors, the crossover operator aims to increase diversity in existing vectors. Achieving further diversification through crossover involves performing an exchange between mutant vector ${v_{i,j}}^k$ and vector ${x_{i,j}}^k$, which results in a test vector ${u_i}^k$, modeled with the following behavior:

$$u^k = u_i^k \begin{cases} v_{i,j}^k, & if j = j_{rand(0,1)} \; or \; rand(0,1) \leq CR \\ x_{i,j}^k, & otherwise \end{cases} \tag{2.10}$$

where $j_{rand(0,\,1)}$ represents one of the positions of the *i*th vector, while *rand(0, 1)* is a random number uniformly distributed between *zero* and *1*.

This operation and random exchange between positions of the population results in a suitable strategy if it is performed with moderate frequency. To regulate the probability that crossover is applied, it is important to calibrate the CR parameter, also called *crossover probability*. Since CR refers to a probability, its range of values goes from *zero* to *1*.

4. **Selection**: This operator has the task of retaining the best solutions found in the mutation and crossover processes. The test vector ${u_i}^k$ may be chosen to belong to the population if it meets the criteria of this operator, detailed in the following equation:

$$x_i^k = \begin{cases} u^k, & if \, f\left(u^k\right) \leq f\left(x_i^k\right) \\ x_i^k, & otherwise \end{cases} \tag{2.11}$$

where *f()* represents the objective function to be optimized. This operation is repeated for each generation until a certain stop criterion is reached.

The differential evolution operators have been widely replicated in the literature due to their simplicity of implementation coupled with their high efficiency in finding suitable solutions to complex optimization problems.

2.3 Algorithms Based on Swarm Intelligence

Among the classifications of metaheuristic algorithms, we have those based on swarm intelligence, i.e., those methods in which collective behavior phenomena are modeled and how the interaction between individuals can generate search patterns that are effective in the implementation of optimization problems [11].

These types of algorithms are mostly based on biological phenomena, bringing inspiration from the functioning of ecosystems, as will be seen below.

2.3.1 *Particle Swarm Optimization (PSO)*

The Particle Swarm Optimization (PSO) algorithm is an algorithm inspired by the collective behavior developed by groups of animals, such as flocks of birds or schools of fish in the sea [12]. This algorithm consists of a swarm of particles or individuals that model this group's behavior. The interaction between them evolves the individual characteristics so that the global strategy can find the best solution.

Each individual has a velocity with which it will be moving through the search space, which will evolve through the iterations as the particle interacts with the others. The motion of each particle is given by two factors: the path traveled by the particle itself and the path traveled by the other particles, referred to as local influence and global influence, respectively [13].

The computational implementation of this algorithm consists of the four steps shown below:

1. **Initialization**: A group of particles is generated at random points in the search space as follows:

$$X_i = lb + R \cdot (ub - lb) \tag{2.12}$$

where X_i represents the *i*th generated individual from a population of n number of particles at iteration k, ub and lb represent the upper and lower bounds, respectively, and R is a uniformly distributed random number.

The parameters c_1, c_2 and v_i^k, which represent the cognitive and social constants, as well as the velocity of the *i*th particle at iteration k are initialized with an initial value of *zero*, respectively.

After this process, the evaluation of the individual in the objective function is performed, where:

$$fX_h = f(X_h) \tag{2.13}$$

2. **Best particle**: The record of the position of the best particle in the population (g) is taken, as well as the record of the best position that the particle itself has had (p).
3. **Velocity**: The new velocity of the particle is calculated based on the following equation:

$$v_i^{k+1} = v_i^k + c_1 \cdot \left(R \cdot \left(p_i^k - x_i^k\right)\right) + c_2 \cdot \left(R \cdot \left(g^k - x_i^k\right)\right) \tag{2.14}$$

where R is a uniformly distributed number and v_i^{k+1} is the new velocity of the *i*th particle.

Each particle defines a new velocity influenced by different factors, such as the global best particle (g_k), the best position of the actual particle (p_i^k) and the current velocity of the particle (v_i^k), as shown in Fig. 2.3.
4. **Particle motion:** Particles move according to the following equation:

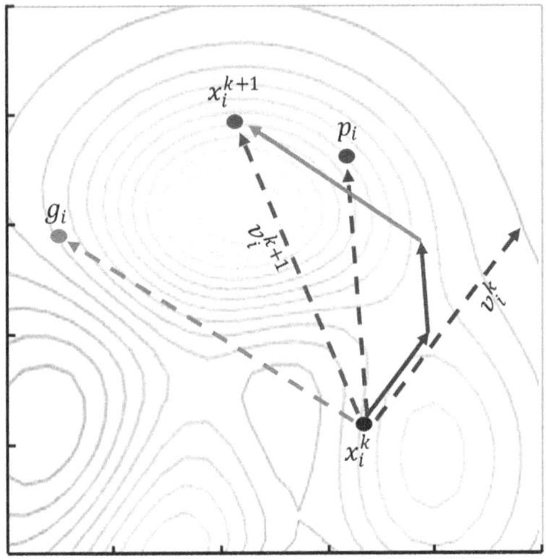

Fig. 2.3 Influences on the motion of a particle

$$x_i^{k+1} = x_i^k + v_i^{k+1} \tag{2.15}$$

where x_i^{k+1} is the new position of the particle. After this process, the evaluation of the individual with respect to the objective function is performed.

Step 1 is only executed when we start the algorithm, and steps 2–4 are executed iteratively until a stopping criterion is reached, which is usually reaching a certain number of iterations.

2.3.2 Bat Algorithm (BA)

The bat algorithm (BA) is based on the echolocation of bats and how they interact with each other to emit pulses and find their prey [14]. To do this, it is important that the bat can identify where there are obstacles and where there could be prey power through its pulses, so for this, the bat must emit a wave with a certain frequency f_i at a certain wavelength λ_i and a sufficient volume level to be heard [15].

Bats communicate over a given range of frequency and wavelength, defined by $[f_{min}, f_{max}]$ and $[\lambda_{min}, \lambda_{max}]$, respectively.

For the application of this algorithm, it is necessary to take into account the following criteria:

1. We define the positions of the bats as x_i^k and their velocities as V_i^k in a search space.
2. We determine the new positions, frequencies and velocities of the bats by the following equations:

$$f_i = f_{min} + (f_{max} - f_{min})\beta \tag{2.16}$$

$$V_i^t = V_i^{t-1} + (x_i^t - x_*)f_i \tag{2.17}$$

$$x_i^t = x_i^{t-1} + V_i^t \tag{2.18}$$

where β is a uniformly distributed random number and x_* is the best location found by the bats so far. Because the product of λ_i by f_i is the increase in bat velocity, one of these parameters functions as a velocity regulator. At the same time, the other may be a fixed factor, depending on the characteristics of the problem at hand.

3. For the local search, once one of the best solutions found is selected, we proceed to calculate a new position of the bat based on the following equation:

$$x_{new} = x_{old} + \varepsilon A^t \tag{2.19}$$

where ε is a uniformly distributed random number while A^t is the average volume of all bats in the population at that instant of time.

4. The parameters A_i and r_i are updated according to the passage of iterations in the algorithm, so that the volume decreases once the bat has found its prey, while the frequency of pulse emission increases, as modeled in the following equation:

$$A_i^{t+1} = \alpha A_i^t,\ r_i^{t+1} = r_i^0[1 - e^{-\gamma t}] \tag{2.20}$$

where α and γ are constants.

Step 1 is performed only when the algorithm starts, while steps 2–4 are executed iteratively until some stopping criterion is reached.

2.3.3 *Moth-Flame Optimization (MFO)*

The Moth-Flame Optimization (MFO) algorithm is a metaheuristic algorithm based on the motion of moths, which in turn is based on transverse orientation. Transverse orientation motion produces the effect in which the moths rotate in a spiral around a light source so that they progressively approach the center of the source as they rotate in a given orbit. It is worth mentioning that the trajectory of the transverse orientation is only applied by the moths when they are close enough to the light source; otherwise, they will maintain a straight trajectory until they enter a sufficiently close area [16].

This algorithm consists of modeling this behavior by taking x as a population of moths exposed to artificial light, where the computational procedure is performed as follows:

1. **Initialization**: each individual is generated at a point in the search space as follows:

$$x_i = lb + R \cdot (ub - lb) \tag{2.21}$$

where x_i represents the generated individual, *ub* and *lb* represent the upper and lower limits, respectively, and *R* is a uniformly distributed random number. Also, a parameter $b = 1$ is initialized, which is a constant that defines the shape of the logarithmic spiral that models the transverse orientation.

After this process, the evaluation of the individual in the objective function is performed, where:

$$fx_i^k = f(x_i^k) \tag{2.22}$$

2. **Ordering**: The moth positions are ordered from best to worst based on their value in the objective function.
3. **Local solutions**: The local solutions *F* in the iteration *k* are initialized based on the best particles:

$$F^k = x_{sorted}^k \tag{2.23}$$

4. **Best solution**: The best overall solution at iteration *k* is recorded as *g*, which will be defined as the light source.
5. **Adjustment of solutions**: The new number of local solutions is calculated by the following equation:

$$nF = int\left(N - k\frac{N-1}{k_{max}}\right) \tag{2.24}$$

where *nF* is the number of local solutions, *N* is the number of moths, *k* is the current iteration number, and k_{max} is the total number of iterations, showing that this algorithm bases its population on the total execution time.
6. **Convergence constant**: The convergence constant a is calculated as shown below:

$$a = \frac{-k}{k_{max}} - 1 \tag{2.25}$$

7. **Transverse orientation**: The new position of the moths is calculated using the transverse orientation as follows:

$$x_i^k = D_i^k e^{br} \cos(2\pi r) + F_j^k \tag{2.26}$$

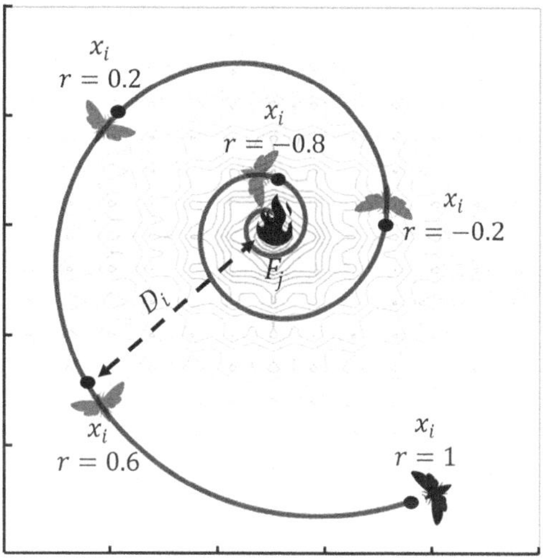

Fig. 2.4 Transverse orientation of a moth with different values for r

where r is a uniformly distributed random number with an interval of *[−2, 1]*, and D_i^k is the absolute value of the distance between the moth x_i^k and the local solution F_j^k.

After this, the moth is re-evaluated in the objective function, and steps 3 and 4 are repeated.

The motion with transverse orientation allows us to explore regions around the flame. This increases the probability of finding better solutions along the way thanks to the decreasing sine-exponential behavior, as shown in Fig. 2.4.

Step 1 is only executed when the algorithm is started, while steps 2–7 are executed iteratively until the stopping criterion is reached, which is given exclusively by reaching the k_{max} number of iterations.

2.3.4 Crow Search Algorithm (CSA)

The Crow Search Algorithm (CSA) is an algorithm based on the collective behavior of crows in which we try to model their movement by storing excess food in their caches so that they can retrieve it later when they need to feed [17].

It is possible to view each crow as an agent that will move through the search space, while the evaluation in the objective function will be the quality of the food within the cache. We should take into account that the crows will behave in two different ways: as a thief when the crow is searching for the food of other crows and as an evader when the victim crow of the thief takes a random trajectory to evade the thief crow, causing the thief crow to get confused and escape with a random trajectory [18].

In order to have a correct balance between the number of robber and evader crows, a *PC* awareness parameter has been defined, as well as a *fl* parameter that defines the magnitude of the movement between the crows.

The computational procedure of the CSA is defined by the steps shown below:

1. **Initialization**: Each crow is generated at a point in the search space as follows:

$$x_i = lb + R \cdot (ub - lb) \tag{2.27}$$

where x_i represents the generated individual, *ub* and *lb* represent the upper and lower limits, respectively, and *R* is a uniformly distributed random number.

Also, the parameters $fl = 2$ and $PC = 0.1$ are initialized as a typical fit given by the author.

After this process, the evaluation of the individual in the objective function is performed, where:

$$fx_i^k = f(x_i^k) \tag{2.28}$$

2. **Posture Adoption**: It is defined whether the crow will act as a thief or evader based on the following approach:

$$x_i^k = \begin{cases} thief, & r_j \geq PC \\ evader, & r_j < PC \end{cases} \tag{2.29}$$

where r_j is a uniformly distributed random number.

3. **Thief or evader**: If the crow adopts the thief posture, then it will have a motion defined by the following equation:

$$x_i^{k+1} = x_i^k + r \cdot fl \cdot (x_j^k - x_i^k) \tag{2.30}$$

where x_i refers to the current crow, x_j to the victim crow and r to a uniformly distributed random number.

If, on the other hand, the crow adopted the evader posture, then its motion will be defined by the following equation:

$$x_i^{k+1} = lb + R \cdot (ub - lb) \tag{2.31}$$

The combination of these two behaviors allows a chase in which the crows can explore part of the search space between them, generating routes with higher probabilities of finding better solutions, as shown in Fig. 2.5.

As can be seen, this algorithm is made up of two different operators adopted by the crows according to a certainly given probability, so we see how when a crow adopts the avoidance posture, then it will tend to have a higher capacity to explore the search space. At the same time, if the crow adopts a thief posture, then it will tend to have a higher capacity to exploit the area between it and the victim crow.

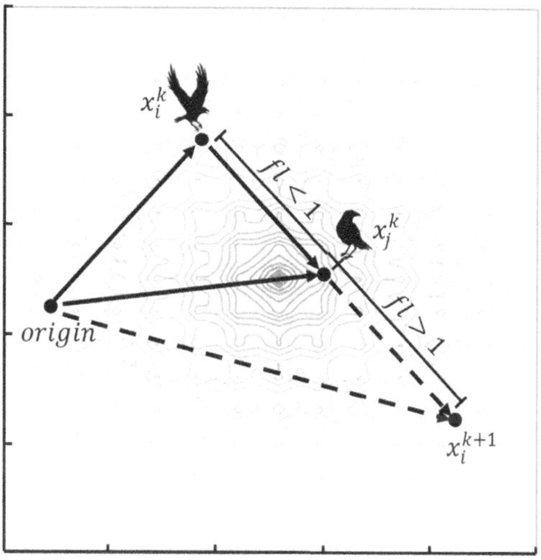

Fig. 2.5 Motion of the thief crow (x_i^k) towards the evader crow (x_j^k) in function of the value of *fl*

From this, we can define the quality of the individuals in finding an optimal solution based on the good handling of the PC parameter, so the simplicity of the operators will perform well if the initial parameters are set appropriately according to the optimization problem to be addressed.

2.3.5 Cuckoo Search (CS)

The Cuckoo Search (CS) algorithm is an algorithm based on the phenomenon of cuckoo parasitism, whereby a female cuckoo lays her eggs in the nest of a bird of another species. When the eggs hatch, the newly hatched cuckoos take care of getting rid of the other eggs in the nest, which are the only ones to be raised by the other bird species [19].

Modeling this behavior toward an optimization method is based on three definitions:

1. Each cuckoo bird can carry one egg at a time and cannot obtain another until depositing the current one in random united.
2. Each egg represents a solution, so nests with better eggs will be more attractive for cuckoos to deposit their eggs in subsequent generations.
3. The host bird can discover whether the egg in the nest is its own or not, so it can decide to either raise the cuckoo or get rid of it.

Within the algorithm, each egg represents an x_i^k position with its respective value in the objective function, so for the placement of new eggs, the cuckoos will fly to new nests using Lévy flights.

The CS computational procedure is defined below:

1. **Initialization**: Each individual is generated at a point in the search space as follows

$$x_i = lb + R \cdot (ub - lb) \tag{2.32}$$

where x_i represents the individual generated, ub and lb represent the upper and lower limits, respectively, and R is a uniformly distributed random number. The parameter $p_a = 0.25$, which is called the discovery probability, is initialized.

After this process, the evaluation of the individual in the objective function is performed, where:

$$fx_i^k = f(x_i^k) \tag{2.33}$$

2. **Mantegna's algorithm to implement Lévy's flight**: It is applied from the implementation of the following equation:

$$s_i = \frac{u}{|v|^{\frac{1}{\beta}}} \tag{2.34}$$

where s_i represents the step size, β is 3/2, and the vectors u and v are defined respectively as:

$$u = N\left(0, \sigma_u^2\right), v = N(0, \sigma_v^2) \tag{2.35}$$

where N is a normally distributed random number with mean *zero* and standard deviation $\sigma^2{}_u$ and $\sigma^2{}_v$, respectively, where:

$$\sigma_u = \left(\frac{\Gamma(1+\beta)\sin\left(\pi\frac{\beta}{2}\right)}{\Gamma\left(\frac{1+\beta}{2}\right)2^{\frac{\beta-1}{2}}\beta}\right)^{\frac{1}{\beta}}, \sigma_v = 1 \tag{2.36}$$

being $\Gamma(\lambda)$ the gamma distribution.

After this, we can finally calculate the Lévy flight by means of the following equation:

$$L_i = 0.01 s_i(x_i^k - x_{best}^k) \tag{2.37}$$

being $x_{best}{}^k$, the best-positioned individual of iteration k.

From Lévy's flight, we can then define the movement of individuals as follows:

$$x_i^{k+1} = x_i^k + L_i \tag{2.38}$$

3. **Replacement:** The cuckoo parasitism simulation is implemented as follows:

$$x_i^{k+1} = \begin{cases} x_i^k + r\left(x_j^k - x_h^k\right), & r1 < p_a \\ x_i^k, & r1 \geq p_a \end{cases} \quad (2.39)$$

where r is a uniformly distributed random number, while $x_j{}^k$ and $x_h{}^k$ are randomly selected individuals from the population.

4. **Elitism:** The best results prevail over time while the worst ones disappear by the following rule:

$$x_i^k = \begin{cases} x_i^{k+1}, & fx_i^{k+1} < fx_i^k \\ x_i^k, & fx_i^{k+1} \geq fx_i^k \end{cases} \quad (2.40)$$

Step 1 is only executed when the algorithm starts, while steps 2–4 are executed iteratively until the stopping criterion is reached.

2.4 Algorithms Based on Physical Laws

There are metaheuristic algorithms that are inspired by physical laws, such as states of matter or harmonics, as well as algorithms based on chemical reactions [20].

2.4.1 Simulated Annealing (SA)

The Simulated Annealing (SA) algorithm is an optimization method based on the annealing of steel and ceramics. This algorithm simulates the heating and cooling process of a material, as well as the variations of its physical properties [21].

For the correct modeling of this physical phenomenon, it is necessary to model a single solution x_k per iteration, so to avoid the problem of stagnation in a local optimum in our single solution, it will be necessary to implement a method so that, if the generated solution does not improve with respect to its previous generation, then by means of a probabilistic variable it can be accepted, with which then we have that the solution will be updated under the following terms [22]:

1. If $f(x^{k+1}) < f(x^k)$
2. If $p_a < r$, where p_a is the probability of acceptance and r is a uniformly distributed random number.

The computational procedure followed by this algorithm is as follows:

1. **Initialization**: The solution x_k is initialized at a random point in the search space, as well as the parameters T_{ini} as the initial temperature of the system, T_{fin} as the final temperature that the system will reach, *Niter* as the maximum number of iterations, $\beta = (T_{ini} - T_{end})/Niter$ as the cooling ratio, σ as the standard deviation and $T = T_{ini}$ as the current temperature of the system.

2. **Perturbation**: A random modification is generated based on the following equation:

$$\Delta x = N(0, \sigma T) \tag{2.41}$$

where N is a normally distributed random number with mean *zero* and standard deviation σT.

3. **Movement**: The position of the solution is updated based on the following equation:

$$x^{k+1} = x^k + \Delta x \tag{2.42}$$

4. **Energy difference**: If the new solution x^{k+1} is worse than the previous solution x^k, then the energy difference is calculated as follows:

$$\Delta f = f\left(x^{k+1}\right) - f\left(x^k\right) \tag{2.43}$$

From this, the probability of acceptance p_a is calculated as follows:

$$p_a = e^{-\frac{\Delta f}{T}} \tag{2.44}$$

5. **Probability of acceptance**: If $p_a < r$, being r a uniformly distributed random number, then $x^{k+1} = x^k$.
6. **Cooling**: The temperature T is updated according to the following model:

$$T = \beta \cdot T \tag{2.45}$$

where due to this, the system tends to cool down and behave differently (Fig. 2.6).

Step 1 is only executed at the start of the algorithm, while steps 2–6 are executed iteratively until the stopping criterion $T = T_{end}$ or $k = Niter$ is reached.

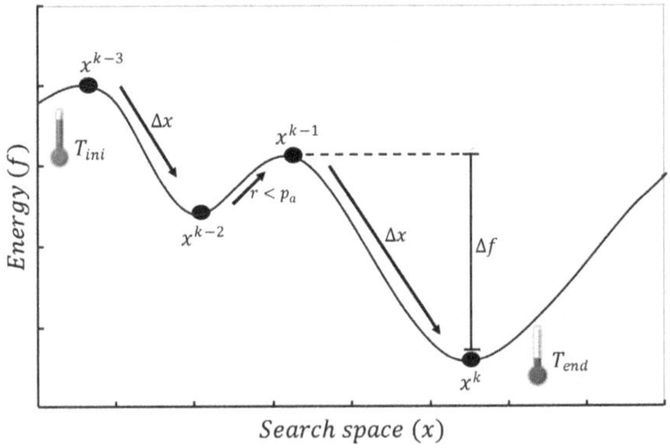

Fig. 2.6 Optimization process using the simulated annealing algorithm

The effectiveness of this algorithm lies in the modeling of the temperature as time passes, causing the solution to progress and its behavior to adapt according to the environment in which it is found.

2.4.2 Harmony Search (HS)

The Harmony Search (HS) algorithm is a metaheuristic method inspired by the process of musical harmonization [23] that searches for an optimal solution by considering various possibilities and iteratively improving them.

The computational implementation of this algorithm is detailed as follows:

1. **Initialization**: the parameters of the algorithm where each solution is generated at a point in the search space are configured as follows:

$$x_i = lb + R \cdot (ub - lb) \tag{2.46}$$

where x_i represents the initial harmonic memory generated, ub and lb represent the upper and lower bounds, respectively, and R is a uniformly distributed random number.

After this process, the evaluation of the solution set is performed on the objective function, where:

$$fx_i^k = f(x_i^k) \tag{2.47}$$

2. **New harmony**: The process of generating a new harmony is called improvisation, where a new harmonic vector x' is generated using the rules of memory consideration, pitch adjustment and random selection, as shown below:

$$x_i' = x_i' + r \cdot bw \tag{2.48}$$

where r is a uniformly distributed random number, and bw is the bandwidth.
3. **Harmonic memory update**: the generated harmony x' replaces the worst harmony stored in the harmonic memory if the value of the solution in the objective function is better than that of the worst harmony.

Step 1 is only executed when the algorithm is started, while steps 2 and 3 are executed iteratively until the stopping criterion is reached.

One of the main advantages of this method is the low sensitivity to the parameters, so it does not require exhaustive tuning in order to perform well, and, unlike some other optimization algorithms, the Harmony Search has a strong global search capability and can avoid getting trapped in local optima.

2.4.3 State-of-Matter Search (SMS)

The State-of-Matter Search (SMS) algorithm is a metaheuristic method inspired by physical principles that describes the molecular motion originated as a result of the thermal energy present in a substance by means of three fundamental states of matter: solid, liquid and gaseous state [24].

In this algorithm, the search agents are modeled as molecules within a search space that will have a motion defined by the parameters ρ as a uniformly distributed interval, α and β as scalar factors of each state and P as a probabilistic factor, where we will divide the algorithm execution in three stages:

1. Gaseous state: makes up 50–60% of the algorithm run time and defines a high-intensity movement of molecules through the search space. At this stage, the typical configuration of the initial parameters is: $\rho = [0.8, 1.0]$, $\alpha = 0.8$, $\beta = 1.0$ and $P = 0.9$.
2. Liquid state: it lasts approximately 30–40% of the algorithm execution time, where the molecules decrease their motion and interaction. In this stage, the typical configuration of the initial parameters is: $\rho = [0.3, 0.6]$, $\alpha = 0.4$, $\beta = 0.7$ and $P = 0.2$.
3. Solid state: making up 10–20% of the algorithm; at this stage, the molecules have a much more stable structure, moving only in a small range by perturbations. At this stage, the typical configuration of the initial parameters is: $\rho = [0.0, 0.1]$, $\alpha = 0.1$, $\beta = 0.5$ and $P = 0.0$ (Fig. 2.7).

The following steps govern the computational procedure of this algorithm:

1. **Initialization**: a group of molecules is generated at random points in the search space as follows:

$$x_i = lb + R \cdot (ub - lb) \tag{2.49}$$

where x_i represents the generated ith individual from a population of n number of molecules at iteration k, ub and lb represent the upper and lower limits, respectively, and R is a uniformly distributed random number.

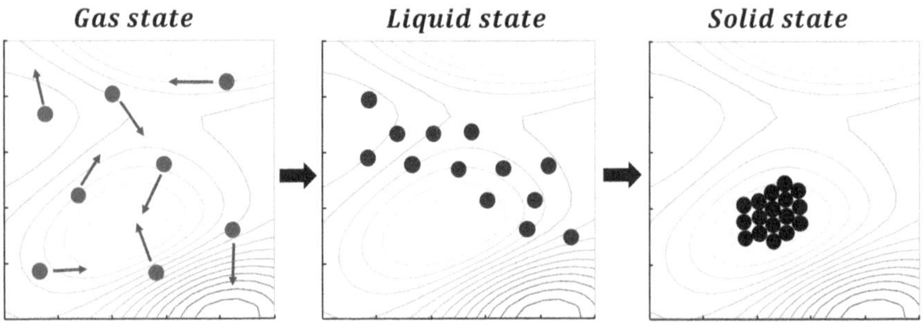

Fig. 2.7 The different states of the SMS algorithm

The parameters ρ, α, β and P are also initialized with the respective values for the gaseous state.

2. **State definition**: Based on the current iteration, the set of parameters to be taken into account is determined according to the defined state of matter.
3. **Molecular motion**: The molecules will move according to the following model:

$$x_i^{k+1} = x_i^k + \rho_{rand} \cdot r \cdot (ub - lb) \cdot v_i^k \tag{2.50}$$

where ρ_{rand} is a random number uniformly distributed within the interval ρ and v_i^t is defined as the velocity of the ith molecule at iteration k defined as:

$$v_i^k = v_{init} \cdot d_i^k \tag{2.51}$$

where v_{init} is the initial velocity of the molecules, defined by the following expression:

$$v_{init} = \beta \frac{\sum_{n=1}^{D} (ub_n - lb_n)}{D} \tag{2.52}$$

Denote ub_n and lb_n as the upper and lower bounds of the nth decision variable of the objective function and D as the number of decision variables for the optimization problem.

On the other hand, $d_i{}^k$ represents a direction vector of the ith molecule defined as:

$$d_i^k = d_i^{k-1} * 0.5\left(1 - \frac{k}{k_{max}}\right) + a_i^k \tag{2.53}$$

where $d_i{}^{k-1}$ represents the previous direction vector and k_{max} the total number of iterations, where $a_i{}^k$ is a unit vector oriented in the direction of the current best solution defined as:

$$a_i^k = \frac{(g_{best} - x_i^k)}{\|g_{best} - x_i^k\| + \varepsilon} \tag{2.54}$$

where g_{best} is the current best solution, ε is a constant of very small value to prevent indeterminacies and $||g_{best} - x_i{}^k||$ is the Euclidean distance between the current molecule and the current best solution.

4. **Molecular collisions**: The perturbation that a molecule undergoes when interacting with other molecules is modeled as follows:

$$if\left(r_{i,j} < Ryi \neq j\right),\ then\, d_i^k = d_j^k y d_j^k = d_i^k \tag{2.55}$$

with $r_{i,j}$ being the Euclidean distance between the colliding particles and R the molecular collision radius defined as:

$$R = \alpha \frac{\sum_{n=1}^{D} (ub_n - lb_n)}{D} \tag{2.56}$$

5. **Random behavior**: The molecules will have the possibility to move through the search space in a random manner defined by the following expression:

$$x_i^{k+1} = \begin{cases} lb + r \cdot (ub - lb), & r \leq P \\ x_i^{k+1}, & r > P \end{cases} \tag{2.57}$$

with r being a uniformly distributed random number.

Step 1 is performed only when the algorithm starts, while steps 2–5 are executed iteratively until the stopping criterion is reached.

2.5 Algorithms Based on Human Behavior

Throughout time, not only have we tried to mathematically model natural phenomena such as the behavior of birds or the physical reactions of the environment, but we have also tried to replicate such complex phenomena as human behavior, their interaction in different circumstances as well as their way of relating to others [25].

Modeling human behavior covers topics such as psychology, sociology, political Science, education and even philosophy.

2.5.1 Small-World Optimization Algorithm (SWOA)

Small World Optimization (SWOA) [26] is a method that seeks to realize the concept known in the world of sociology as the "small world." The small world theory reflects the closeness of human connections, emphasizing a global human interconnectedness where there is a large network of communication through common nodes.

The SWOA optimization strategy consists of the management of two types of clusters: high clustering and short clustering. High clustering corresponds to a global search strategy, prioritizing exploration within the search space, while short clustering focuses on an exploitation strategy within promising regions. The computational procedure of this algorithm is detailed in the following ** steps:

1. **Initialization**: A group of individuals is generated at random points in the search space as follows:

$$x_i = lb + R \cdot (ub - lb) \tag{2.58}$$

where x_i represents the generated ith individual from a population of n number of individuals at iteration k, ub and lb represent the upper and lower limits, respectively, and R is a uniformly distributed random number.

The parameters p and k are also initialized based on the probability of creating random connections during the construction of the network and the number of neighbors that define the initial connectivity of the network, respectively.

2. **Small-world network construction**: When constructing the small-world network, a base structure is created using the Watts-Strogatz model, which ensures a combination of global and local connections, considering the following criteria:

- Each node x_i is connected to its k nearest neighbors.
- Each local connection is replaced with a probability p by a connection to another randomly selected node.

3. **Local Exploitation**: In this stage, the position of individual x_i is updated by means of the information provided by its local neighbors; this phenomenon is modeled by means of the following equation:

$$x_i^{t+1} = x_i^t + \alpha \cdot \sum_{j \in N_{local}} \left(x_j^t - x_i^t\right) + \beta \cdot R \tag{2.59}$$

where α is a factor limiting the influence of the individual's near neighbors, β is a noise scaling factor, and R is a uniformly distributed random number ranging from -1 to 1.

4. **Global Exploration**: If individual x_i has connections to distant nodes, then an additional operation is implemented to increase the exploration capability across the search space and enhance diversity:

$$x_i^{t+1} = x_i^t + \gamma \cdot \left(x_j^t - x_i^t\right) \tag{2.60}$$

where x_j is a randomly connected global node, and γ is an influence factor on global connections.

5. **Selection**: This operation has the job of keeping the best solutions within the network, while solutions that do not meet a certain standard are discarded. This criterion is detailed in the following equation:

$$x_i^{t+1} = \begin{cases} x_i^{t+1}, & \textit{if } f\left(x_i^{t+1}\right) \leq f\left(x_i^t\right) \\ x_i^t, & \textit{otherwise} \end{cases} \tag{2.61}$$

where $f()$ represents the objective function to be optimized. This operation is repeated for each generation until a certain stop criterion is reached.

6. **Network adaptation**: the network can dynamically adapt its structure based on the modification of its essential parameters:

- The network can reduce the number of global connections over time, implying that $p \rightarrow 0$.
- The network can adjust the number of neighbors k to modify the dynamics of the interaction between individuals in the population.

Step 1 is only executed at the start of the algorithm, while steps 2–5 are repeated until the algorithm reaches a certain stopping criterion.

2.5.2 Teaching-Learning-Based Optimization (TLBO)

Within the optimization methods, Teaching-Learning-Based Optimization (TLBO) [27] is an algorithm that has had a certain impact thanks to the simplicity of its operators together with its structure, which favors the exchange of information between its individuals. This technique bases its operation on the teaching process that exists between a teacher, and a group of students called "learners," who will learn from the information given by the individual teacher in order to improve their statistics and achieve better performance. This algorithm also contemplates the learning process that happens between the students, where they exchange information with the objective of improving together, acquiring new information and evaluating whether to replace the current information with the information of the other individuals or to stay with the current information, rejecting the data of the others if necessary.

The computational implementation of this algorithm is quite simple and can be summarized in 3 defined steps:

1. **Initialization**: A group of "learners" is generated at random points in the search space as follows:

$$x_i = lb + R \cdot (ub - lb) \tag{2.62}$$

where x_i represents the generated ith learner from a population of n number of individuals at iteration k, ub and lb represent the upper and lower limits, respectively, and R is a uniformly distributed random number.

2. **Teacher Phase**: This phase simulates learning in a group of learners, where the teacher $x_{teacher}$ is the best-evaluated individual in the population. The teacher will be in charge of transferring information to the rest of the population, so his influence will be reflected in a linear movement within the search space, modeled by the following equation.

$$x_i^{k+1} = x_i^k + r \cdot (x_{teacher} - T_F \cdot M_i) \tag{2.63}$$

where x_i^{k+1} is the new solution for the ith learner, r is a random number uniformly distributed between *zero* and *1*, $x_{teacher}$ is the position of the individual "teacher," T_F is the Teaching Factor, which can be *zero* or *1* randomly selected. M_i is the average of all current learners.

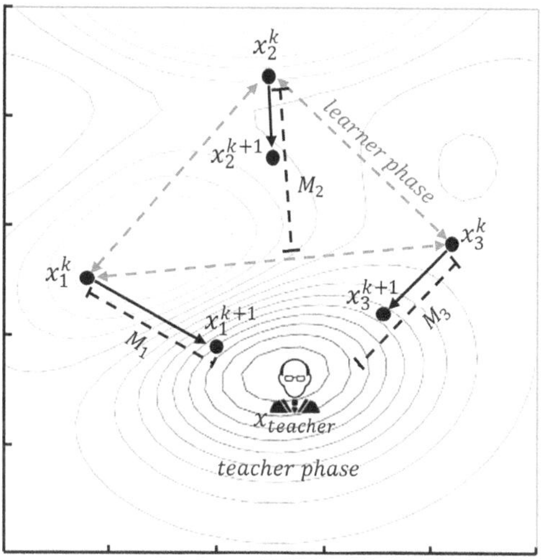

Fig. 2.8 Motion made by learners (x_1, x_2 and x_3) in the teacher stage and learner stage

3. **Learner Phase:** In this stage, trainees increase their performance by sharing information. During this phase, each trainee will interact with another trainee randomly to compare their fitness so that the learner with poorer fitness will acquire information from the learner with better fitness, performing a movement as modeled below:

$$x_i^{k+1} = \begin{cases} x_i^k + r \cdot \left(x_i^k - x_j^k\right), f(x_i^k) < f(x_j^k) \\ x_i^k + r \cdot \left(x_j^k - x_i^k\right), \quad otherwise \end{cases} \tag{2.64}$$

where $x_i{}^k$ is the current trainee and $x_j{}^k$ is the trainee randomly selected from the population (Fig. 2.8).

While step 1 is only executed at the start of the algorithm, steps 2 and 3 are repeated until the algorithm reaches a stopping criterion.

References

1. M. Abdel-Basset, L. Abdel-Fatah, A. K. Sangaiah, Metaheuristic algorithms: a comprehensive review, in *Computational Intelligence for Multimedia Big Data on the Cloud with Engineering Applications* (Elsevier, 2018), pp. 185–231. https://doi.org/10.1016/B978-0-12-813314-9.00010-4
2. T. Dokeroglu, E. Sevinc, T. Kucukyilmaz, A. Cosar, A survey on new generation metaheuristic algorithms. Comput. Ind. Eng. **137**, 106040 (2019). https://doi.org/10.1016/j.cie.2019.106040
3. S.S. Vinod Chandra, H.S. Anand, Nature inspired meta heuristic algorithms for optimization problems. Computing **104**(2), 251–269 (2022). https://doi.org/10.1007/s00607-021-00955-5
4. T. Bäck, H.-P. Schwefel, An overview of evolutionary algorithms for parameter optimization. Evol. Comput. **1**(1), 1–23 (1993). https://doi.org/10.1162/evco.1993.1.1.1

5. T. Bartz-Beielstein, J. Branke, J. Mehnen, O. Mersmann, Evolutionary algorithms. WIREs Data Min. Knowl. Discov. **4**(3), 178–195 (2014). https://doi.org/10.1002/widm.1124
6. S. Katoch, S.S. Chauhan, V. Kumar, A review on genetic algorithm: past, present, and future. Multimed. Tools Appl. **80**(5), 8091–8126 (2021). https://doi.org/10.1007/s11042-020-10139-6
7. Y.C. Li, L.N. Zhao, S.J. Zhou, Review of genetic algorithm. Adv Mat Res **179–180**, 365–367 (2011). https://doi.org/10.4028/www.scientific.net/AMR.179-180.365
8. H.-G. Beyer, *The Theory of Evolution Strategies*, vol. 1 (Springer Science & Business Media, 2013)
9. H.-G. Beyer, H.-P. Schwefel, Evolution strategies: a comprehensive introduction. Nat. Comput. **1**(1), 3–52 (2002). https://doi.org/10.1023/A:1015059928466
10. R. Storn, K. Price, Differential evolution – a simple and efficient heuristic for global optimization over continuous spaces. J. Glob. Optim. **11**(4), 341–359 (1997). https://doi.org/10.1023/A:1008202821328
11. J. Kennedy, Swarm intelligence, in *Handbook of Nature-Inspired and Innovative Computing* (Kluwer Academic Publishers, Boston), pp. 187–219. https://doi.org/10.1007/0-387-27705-6_6
12. J. Kennedy, R. Eberhart, Particle swarm optimization, in *Proceedings of ICNN'95 - International Conference on Neural Networks*, IEEE, pp. 1942–1948. https://doi.org/10.1109/ICNN.1995.488968
13. D. Wang, D. Tan, L. Liu, Particle swarm optimization algorithm: an overview. Soft. Comput. **22**(2), 387–408 (2018). https://doi.org/10.1007/s00500-016-2474-6
14. X.S. Yang, X. He, Bat algorithm: literature review and applications. Int. J. Bio-Inspired Comput. **5**(3), 141 (2013). https://doi.org/10.1504/IJBIC.2013.055093
15. X. Yang, A. Hossein Gandomi, Bat algorithm: a novel approach for global engineering optimization. Eng. Comput. (Swansea) **29**(5), 464–483 (2012). https://doi.org/10.1108/02644401211235834
16. S. Mirjalili, Moth-flame optimization algorithm: a novel nature-inspired heuristic paradigm. Knowl. Based Syst. **89**, 228–249 (2015). https://doi.org/10.1016/j.knosys.2015.07.006
17. A. Askarzadeh, A novel metaheuristic method for solving constrained engineering optimization problems: crow search algorithm. Comput. Struct. **169**, 1–12 (2016). https://doi.org/10.1016/j.compstruc.2016.03.001
18. A.G. Hussien et al., Crow search algorithm: theory, recent advances, and applications. IEEE Access **8**, 173548–173565 (2020). https://doi.org/10.1109/ACCESS.2020.3024108
19. X.-S. Yang, S. Deb, Cuckoo search via Lévy flights, in *2009 World Congress on Nature & Biologically Inspired Computing (NaBIC)*, IEEE (2009), pp. 210–214. https://doi.org/10.1109/NABIC.2009.5393690
20. D. Demirol, F. Oztemiz, A. Karci, Performance comparison of physics based meta-heuristic optimization algorithms. in *2018 International Conference on Artificial Intelligence and Data Processing (IDAP)*, IEEE, September 2018 (2018), pp. 1–5. https://doi.org/10.1109/IDAP.2018.8620742
21. S. Kirkpatrick, C.D. Gelatt, M.P. Vecchi, Optimization by simulated annealing. Science (1979) **220**(4598), 671–680 (1983). https://doi.org/10.1126/science.220.4598.671
22. B. Hajek, A tutorial survey of theory and applications of simulated annealing, in *1985 24th IEEE Conference on Decision and Control*, IEEE, December 1985 (1985), pp. 755–760. https://doi.org/10.1109/CDC.1985.268599
23. Z.W. Geem, J.H. Kim, G.V. Loganathan, A new heuristic optimization algorithm: harmony search. Simulation **76**(2), 60–68 (2001). https://doi.org/10.1177/003754970107600201
24. E. Cuevas, A. Echavarría, M.A. Ramírez-Ortegón, An optimization algorithm inspired by the states of matter that improves the balance between exploration and exploitation. Appl. Intell. **40**(2), 256–272 (2014). https://doi.org/10.1007/s10489-013-0458-0
25. S.-A. Ahmadi, Human behavior-based optimization: a novel metaheuristic approach to solve complex optimization problems. Neural Comput. Appl. **28**(S1), 233–244 (2017). https://doi.org/10.1007/s00521-016-2334-4

26. H. Du, X. Wu, J. Zhuang, *Small-World Optimization Algorithm for Function Optimization* (2006), pp. 264–273. https://doi.org/10.1007/11881223_33
27. R.V. Rao, V.J. Savsani, D.P. Vakharia, Teaching-learning-based optimization: a novel method for constrained mechanical design optimization problems. Comput. Aided Des. **43**(3), 303–315 (2011). https://doi.org/10.1016/j.cad.2010.12.015

Chapter 3
Population Initialization Based on the Gibbs Sampling Methodology

3.1 Introduction

Metaheuristic algorithms present diverse characteristics that make them capable of solving complex optimization problems efficiently, overcoming diverse limitations of classical optimization approaches, placing metaheuristics as widely efficient alternatives for solving high-complexity non-linear problems [1] applied in fields of engineering, economics, computer sciences, and other applications [2, 3]. Metaheuristic algorithms are populational; that is to say, they employ a set of different elements denoted as search agents distributed along a search space determined by the nature of each optimization problem. The search agents have the main objective of exploring different sections with movement determined by the evolutionary operators, aiming to find positions with potential quality until the global solution is found [4].

Population initialization is the first step in the evolutionary process of metaheuristic algorithms and plays an important role in the evolutionary process of metaheuristic algorithms [5]. In this stage, the positions for the initial set of candidate solutions are configured, and afterward, the algorithm is employed to consolidate its optimization strategy. The initialization plays an important role in the performance of algorithms since having an adequate distribution of the search agents where the potential regions are easily detected given the quality of solutions facilitates the task of searching for the algorithm by identifying regions with potential solutions from early stages of the algorithm's process [6]. Despite the advantages and enhancement of the metaheuristic algorithms capabilities provided by a proper setting, the research field of population initialization presents a high potential for the improvement of the efficiency of classic and novel metaheuristic schemes not fully explored since researchers tend to focus their efforts on the development of novel searching operators or improving the classical evolutionary operators.

In the literature, the population initialization techniques are classified as deterministic and stochastic based on the degree of randomness for the generation of

E. Cuevas et al., *Advanced Metaheuristics: Novel Approaches for Complex Problem Solving*, Studies in Computational Intelligence 1210,
https://doi.org/10.1007/978-3-031-89284-4_3

variables constituting initial solutions. Deterministic initialization methods are characterized by their limited dependence on the generation of random numbers but also focus on generating samples uniformly distributed along the space, resulting in similar populations in different executions. Some examples of deterministic initialization methods include the Halton method [7], the Latin Hypercube method [8], the "Good lattice points" method [9] and the Diagonal Linear Uniform method [10]. Conversely, stochastic methodologies depend entirely on the generation of random variables for setting the initial solutions, thus generating a unique set of solutions per execution. Stochastic methods often involve probabilistic distributions or positions opposite to those previously generated. Some examples of stochastic initialization techniques are the Opposition-Based Learning (OBL) method [11], the Chaotic maps method [12], the Adaptive Randomness [13], and a recent method employing the Metropolis–Hastings sampling technique [14].

Each initialization method affords a unique focus for setting a group of candidate solutions within a determined space, often combining stochastic techniques to create unique solutions through probabilistic criteria alongside deterministic methods to maintain a more controlled distribution [15]. These initialization methodologies introduce different properties that potentialize the performance of the optimization algorithms, as the approach of the individuals to promising regions by considering not only a spatial distribution but the fitness of the initial solutions, driving a better exploration of the space by easily recognizing locations with poor performance.

However, recent studies have evidenced a common issue with preserving a uniform distribution present in different initialization methodologies when facing high-dimensional optimization challenges [16]. As the number of decision variables increases for the individuals of the population, several initialization mechanisms present issues, such as the grouping of the individual in determined regions of the search space, considerably restricting the exploration of the environment and the capability of metaheuristic algorithms to reach optimal solutions. This issue is given by the nature of the generation of random numbers, which, by generating a greater number of samples, the random numbers tend to cluster the samples in determined values. Employing mathematical and probabilistic models for generating samples has proven to be a more effective method for initializing populations with high performance, even in high-dimensional scenarios, giving as an example the Metropolis–Hastings method, which employs a probabilistic selection based on the quality of the solution for the generation of samples with a low degree of correlation between the existing individuals. Employing methods with a minor degree of correlation between the generated individuals may guarantee a better distribution, regardless of the dimensionality of the optimization problem, ensuring further better coverage of the search space.

Gibbs Sampling [17] is a statistical methodology applied for the sampling of multivariate probabilistic distributions. This sampling method is particularly useful for distributions with difficult direct sampling. This method is applied in high-dimensional spaces where each variable corresponds to an element in the distribution. After the individual is generated, this method follows a sequential sampling for a new individual. For each sequence, a new sample is generated for a new decision variable, while the other generated samples are fixed. This sampling algorithm is looped through all the decision variables, generating new values once per loop, considering only the previous sample.

This chapter introduces a novel initialization methodology for metaheuristic algorithms employing the Gibbs Sampling methodology. This initialization employs the Gibbs sampling for the generation of decision variables for each search agent based on a multidimensional Gaussian distribution aiming to get better coverage of the solution space denoted by a cost function. This method generates samples sequentially by generating one single sample at a time, each decision being completely independent of the other previously generated sample. This sampling method aims for a low degree of correlation between the decision variables, maintaining a better separation of the search agents. This feature represents a major advantage for solving high-dimensional optimization problems, given the better dispersion of the solutions along a search space and the prevention of premature convergences. For a proper validation of this novel initialization, it has been embedded in a traditional Differential Evolution (DE) algorithm and compared employing five different initialization methods reported in the literature through a set of 30 test functions, selected by their diversity of complexity and challenging landscapes. The numerical results demonstrate a significative capability from the initialization technique to potentialize the search mechanisms from the DE algorithm, aiding in finding the global optimal solution more efficiently and requiring a minor number of evaluations in the cost function.

This chapter is organized as follows: Sect. 3.2 discusses the impact of the initialization in the metaheuristic algorithms. Section 3.3 classifies the initialization techniques and analyzes diverse examples reported in the literature, Sect. 3.4 discusses different issues existing in different initialization approaches, Sect. 3.5 presents a detailed description of the Gibbs sampling, Sect. 3.6 details the general procedure of the Gibbs initialization method, Sect. 3.7 analyzes the experimental results, and finally, Sect. 3.8 discusses the conclusions.

3.2 Population Initialization

For all populational metaheuristic algorithms, a set of candidate solutions must be defined. This process plays an important role, as effective strategies from metaheuristic algorithms can be put in place early on to locate high-quality solutions [5]. This stage presents a high relevance to determining the exploration and exploitation capabilities of the algorithm, avoiding premature convergence in suboptimal regions or an inefficient exploration process, potentializing the evolutionary process of the algorithm considerably, aiding in getting high-quality solutions for diverse, complex real-life optimization problems, by considering rather a proper spatial distribution covering different places in the solution space or placing the search agents strategically by the afforded solutions in an early stage of the algorithm [6].

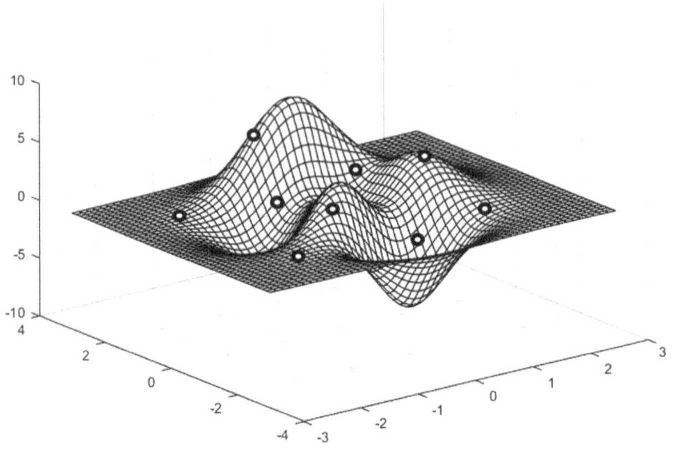

Fig. 3.1 Graphical representation of an initial configuration of candidate solutions

Generally, most metaheuristic algorithms employ a random uniform initialization, which generates values for the decision variables of each search agent considering a uniform distribution delimitated by the lower and upper boundaries of the solution space. This initialization process is defined as follows in Eq. (3.1). In this Equation, $\mathbf{x}_{i,j}$ corresponds to the j th decision variable for the i th search agent of the population.

$$\mathbf{x}_{i(1,\ldots,n),j(1,\ldots,d)} = \boldsymbol{P}_{lb,j} + U(0,1)\cdot\left(\boldsymbol{P}_{ub,j} - \boldsymbol{P}_{lb,j}\right) \tag{3.1}$$

Consider the representation provided in Fig. 3.1, where an initial configuration of candidate solutions is generated following Eq. (3.1) along a search space determined for the "peaks" cost function, mathematically defined in Eq. (3.2) delimitated with the boundaries $-3.2 \le x \le 3.2$. The placement of the individuals in different places affords a great knowledge of the landscape and facilitates a better decision-making process about which locations are worthy or not to explore.

$$f(x, y) = 3(1-x)^2 + 10\left(\frac{x}{5} - x^3 - y^5\right)e^{\left(-x^2-y^2\right)} - \frac{1}{3}e^{\left(-(x+1)^2-y^2\right)} \tag{3.2}$$

3.3 Classification of the Initialization Techniques

Based on the process performed for generating the initial variables, the initialization approaches may be classified as deterministic or stochastic [5]. This classification is based on the degree of randomness employed to generate the set of decision variables for all the search agents. Each technique affords a different approach for initializing the spaces, providing a better variability from the stochastic methods, as a more uniform distribution and control by the deterministic methods.

3.3.1 *Deterministic Techniques*

Deterministic initialization methods are characterized by their limited dependence on the stochastic methods. Instead, they focus on generating diverse solutions in determined regions of the solution space, which prove to be of promising quality. In general, deterministic initialization methods employ better spatial distribution among the candidate solutions, obtaining better dispersion among the search agents and avoiding premature convergences. Some examples of deterministic initialization methods are the Halton initialization [7], the Latin Hypercube initialization [8], the Good-Lattice Points [9] initialization and the Diagonal Linear Uniform initialization [10].

3.3.1.1 Halton Initialization

This initialization technique was originally proposed by Digehsara et al. [7] and implemented for diverse optimization problems integrated into the Particle Swarm Optimization (PSO) algorithm. This initialization technique follows a scrambled Halton sequence, presenting the generation of populations with a minor degree of correlation between the individuals, an important feature for the proper distribution of the individuals when the number of decision variables increases for the optimization problem. The scrambled Halton sequence utilizes a combination of the prime numbers employed for the generation of the basic Halton sequence [11, 18] to generate a minor correlation between the individuals, avoiding groups of individuals in determined regions of the space. The scrambled proposed Halton sequence is determined as:

$$\varphi_r(g) = \sum_{l=1}^{L} \sigma_r(b_l(g))r^{-l-1} \tag{3.3}$$

In Eq. 3.3, $\varphi(g) = \left(\varphi_{r_1}(g), \varphi_{r_2}(g), \ldots, \varphi_{r_s}(g) in|0, 1\right)^s$ corresponds to an extension to the Van der Corpur sequence proposed by Halton in a s-dimensional space [19], $(r_1, r_2, \ldots, r_s)$ corresponds to a series of prime numbers and $b_l(g)$ corresponds to the base number where $0 \leq b_l(g) \leq r$. The scrambled Halton sequence presents a major advantage for the dispersion of the individuals by considering a shuffle of the r prime numbers to minorize the correlation among the coordinates of the normal Halton sequence. In Eq. (3.2), σ_r corresponds to the combination operator on the

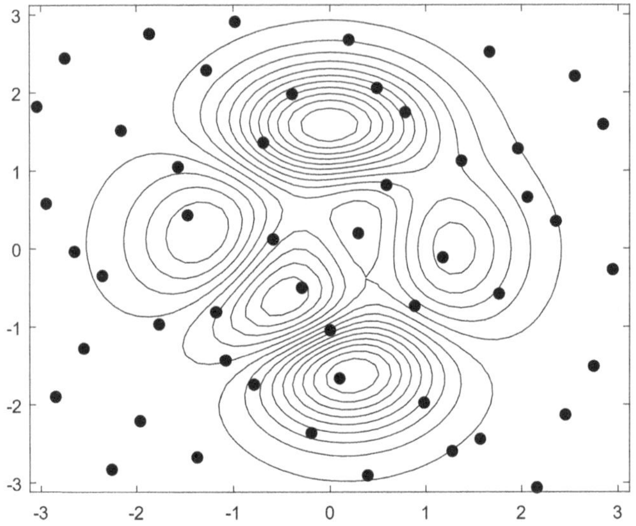

Fig. 3.2 Initial configuration produced by a Halton initialization

numbers of $b_l(g)$. Figure 3.2 presents a graphical example produced with the scrambled Halton sequence initialization applied to a space delimited as $-3.2 \leq x \leq 3.2$ for the function explained in Eq. (3.1).

3.3.1.2 Latin Hypercube Initialization

The Latin Hypercube initialization proposed originally by Mousavirad et al. [8] employs the principles of the Latin Hypercube Sampling for a proper distribution of multidimensional individuals along high-scale searching spaces [20]. The Latin Hypercube Sampling (LHS) technique presents significant advantages for the sampling of multidimensional arrays, keeping a uniform distribution. Mousavirad presented a modification of the traditional LHS sampling technique, placing each sample in the center of the distribution, placing all the samples equally distant in a determined grid of N elements.

Figure 3.3 presents a graphical representation of the initial distribution of candidate solutions employing a Latin Hypercube Sampling (LHS) methodology in a space delimitated as $-3.2 \leq x \leq 3.2$ for the Peaks function established in Eq. (3.1).

3.3.1.3 Good-Lattice Points Initialization

A Good-Lattice points [21] initialization methodology follows the principles of the Good-Lattice rules capable of generating a set of values in a determined range with proper space-filling properties. The Good-Lattice Points (GLP) sets are commonly

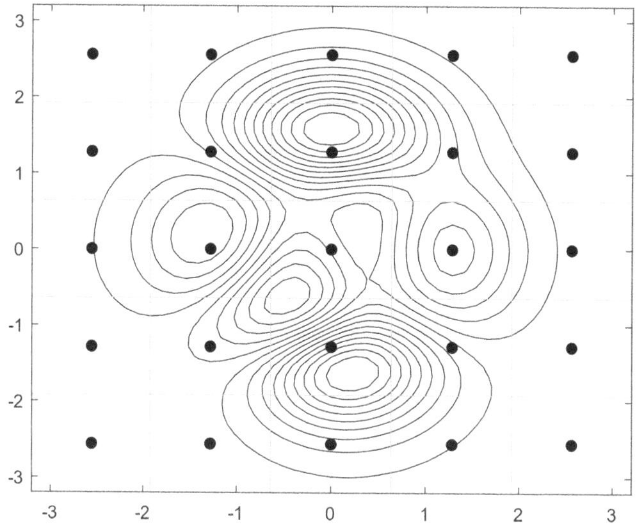

Fig. 3.3 Initial configuration produced by a center-based Latin Hypercube Sampling (LHS) initialization

used in different applications and different models, such as multidimensional quadratures or quasi-Monte Carlo sampling techniques. The uniform Design (UD) [22] models applied for the GLP sets offer better coverage in the space.

Figure 3.4 presents a graphical representation of an initial set of candidate solutions uniformly distributed following the Good Lattice Points rules applied in a space delimitated as $-3.2 \leq x \leq 3.2$ for a Peaks function described in the Eq. (3.1). As can be seen, the individuals present a uniform distribution along the space following equally distributed distances among the individuals.

3.3.1.4 Diagonal Linear Uniform Initialization

The Diagonal Linear Uniform [10] initialization proposed by Li et al. follows a distribution of the individuals across a determined space following a Diagonal Uniform proposed by Chou et al. [23] which claims that dividing the searching space in n subspaces and generating the individuals across the subspaces located in the diagonal presents a better coverage of the space and prevents the clustering of individuals in determined spaces.

Li et al. followed this principle to adjust the location of the individuals in the diagonal spaces and adjust the distances between the individuals following Eq. (3.3):

$$dx = \frac{x_{i\_max} - \overline{x}_{i\_min}}{N - 1} \quad (3.4)$$

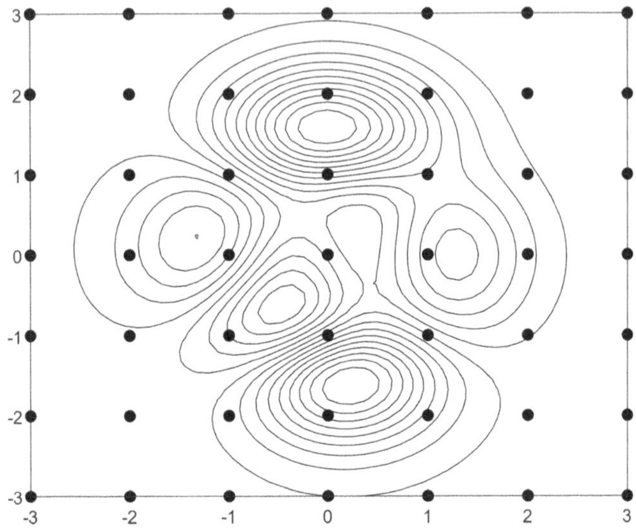

Fig. 3.4 Initial configuration produced by a good-lattice points initialization

where x_{i_max} and $\overline{x}_{i\_min}$ correspond to the upper and lower boundaries for each x_i decision variable and N represents the population size. This initialization method utilizes the diagonal spaces aligned to the vertices of the subspaces. Figure 3.5 shows a graphical representation of a 2-D initial configuration in a space delimitated as $-3.2 \leq x \leq 3.2$ by the "peaks" function established in Eq. (3.1) following the Diagonal Linear Uniform initialization (DLU) technique.

3.3.2 *Stochastic Techniques*

Stochastic initializations are techniques that, contrary to the deterministic methods, stochastic initialization methods are entirely dependent on random mechanisms for generating an initial population. Stochastic methods often employ probabilistic distributions or consider the concept of Opposition in the generation of values. These techniques are easily applicable and offer great coverage of the search space with a minor number of operations. Among the stochastic initialization techniques, some interesting examples are Opposition-Based initialization (OP) [11, 12, 24], Adaptive Randomness initialization (AR) [13], Chaotic Maps initialization (CM) [12] and Metropolis–Hastings initialization (MH) [14].

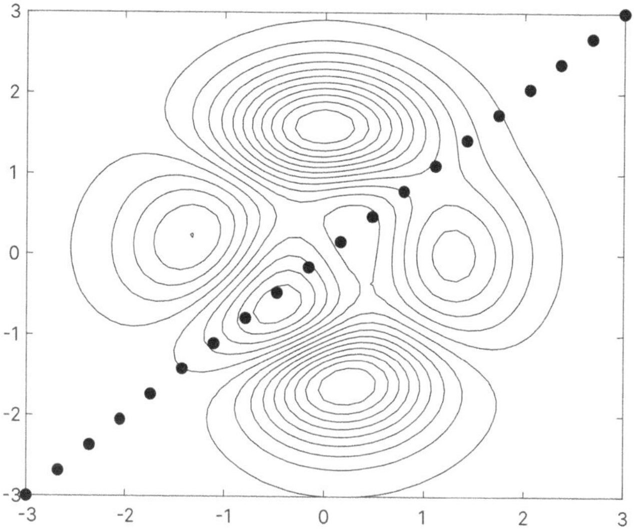

Fig. 3.5 Initial configuration produced by a Diagonal Linear Uniform initialization

3.3.2.1 Opposition-Based Initialization

The Opposition-Based initialization (OP) initialization proposed by Rahnamayan et al. [11] is a population initialization technique that applies the concept of the opposite optimization for the generation of diverse candidate solutions. The Opposition-Based Learning scheme [25] establishes that the consideration of a new position given by the opposite guess can afford a wider knowledge of a given problem, increasing the probability of finding a solution closer to the global optimal solution. The concept of the Opposition established, "*Let x be a real number in an interval* $[a, b](x \in [a, b])$" the opposite number $\check{x}$ is defined by the Eq. (3.5):

$$\check{x}_i = a_i + b_i - x_i \tag{3.5}$$

In this initialization scheme, the population generates two different positions for each candidate solution: One of them is generated randomly following a uniform distribution as in Eq. (3.2), and the other one considers the opposite position in the space, and the position prevailing for the population is the one presenting the best fitness.

3.3.2.2 Adaptive Randomness Initialization

The adaptive randomness (AR) initialization methodology [13] proposed by Pan et al. is a modification of the traditional random initialization to address the issue of the lower distribution of individuals in high-dimensional spaces. The pseudorandom number is often a simulation of the random processes for generating new values. However, when high-dimensional values tend to be generated following these pseudorandom factors, they tend to group or generate similar values.

The Adaptive Randomness methodology incorporates control of the distances between the individuals, ensuring that, when a new individual is generated in the space, the modification ensures a new further individual does not approach a previously generated individual. In this initialization, the calculation of distance between two pairs of individuals follows a Euclidean Distance [16, 26] calculation for al the $t(1, \ldots, d)$ decision variables, defined in Eq. (3.6):

$$d\left(\mathbf{x}_i, \mathbf{y}_j\right) = \sqrt{\sum_{t=1}^{d} \left(\mathbf{x}_{i,t} - \mathbf{y}_{j,t}\right)^2} \tag{3.6}$$

When a new candidate individual is generated, the criterion for being added to the $\boldsymbol{P}$ initial population of n individuals compared the distance between the new individual and the individuals already belonging to the population by following the criteria explained in Eq. (3.7):

$$\min_{1 \leq i \leq k} d(\mathbf{x}_i, \mathbf{y}_{New}) \geq \min_{1 \leq i \leq k} d\left(\mathbf{x}_i, \mathbf{y}_j\right) \tag{3.7}$$

where k is a counter for the n individuals in the population, this criterion ensures the proper distribution of the individuals by considering that a proper distance between other individuals exists among the other members of the population.

3.3.2.3 Chaotic Maps Initialization

The Chaotic Maps Initialization [12, 24] proposed by Ozer is an initialization methodology capable of generating numbers following different chaotic maps instead of generating random numbers, with the objective of reducing the aggregation of individuals in determined regions and enhancing the convergence rate of a metaheuristic algorithm. The objective of employing chaotic maps is to provide better unpredictability in the generation of the decision variables, providing a better spread of the individuals along the space and preventing a premature convergence of minimizing the risk of the individuals getting stuck in local optimal solutions.

In mathematics, the concept of chaos represents a deterministic method of a random process, presenting a random, irregular and unpredictable nature. Chaotic maps are very used in the field of optimization algorithms because of the easy generation of values with a determined degree of randomness, employing easy calculation with a minor number of initial parameters, generating different sequences with a low correlation, and replacing most of the random processes.

This initialization method employs seven different chaotic maps: the logistics maps, the circle map, the Gaussian map, the henom map, the sinus map, the sinusoidal iterator, and the tend m [27].

The chaotic map employed to generate the decision variables for the new individual is further employed for generating the new individual distributing this new position across the search space following the criterion exposed in Equation:

$$\mathbf{x}_{i,j} = \boldsymbol{P}_{lb,j} + \boldsymbol{cm}_{i,j} \times \left(\boldsymbol{P}_{ub,j} - \boldsymbol{P}_{lb,j}\right) \tag{3.8}$$

As can be seen, Eq. (3.8) conserves similarities with the random initialization of Eq. (3.2), replacing the random uniform distributed number with the chaotic map employed for the generation of the values for the decision variables.

3.3.2.4 Metropolis–Hastings Initialization

The Metropolis–Hastings (MH) initialization proposed by Cuevas et al. [14] in a novel initialization technology employing the criterion of the Monte Carlo Markov Chains (MCMC) Metropolis–Hastings acceptance criterion. The Metropolis–Hastings methodology is used for the sampling of values in multidimensional distributions by employing a probabilistic selection where it can be decided whether a candidate sample is accepted for the next generation or discarded. The initialization approach proposed in [14] suggests a probabilistic criterion for the acceptance of new individuals based on the quality of solutions reported by different candidate solutions. To get the probability of new individuals being selected in the population, the fitness values for a previous sample and the current newly generated sample are compared by applying the following criterion:

$$\alpha = \min\left(\frac{J(\mathbf{y})}{J(\mathbf{x}_{i-1})}, 1\right) \tag{3.9}$$

In Eq. (3.9), $J(\mathbf{y})$ represents the result given by the newly generated sample evaluated in the cost function, and $J(\mathbf{x}_{i-1})$ if the result of the evaluation in the cost function for the previous sample in the population. The new candidate solution $\mathbf{y}$ is generated in sequence considering a previous sample from the population, defined in Eq. (3.10):

$$\mathbf{y} = \mathbf{x}_{i-1} + N(0,1) \tag{3.10}$$

where $N(0,1)$ corresponds to a random number from a normalized distribution.

Finally, the acceptance of the new $\mathbf{y}$ individual is considered by a probabilistic selection with a random number from a normalized distribution from 0 to 1. As the difference in the solutions increases, the probability of being accepted increases considerably, accepting most solutions presenting a high quality in the solutions.

3.4 Issues with High-Dimensional Population Initialization

Despite the reported advantages and notorious improvement of the evolutionary capabilities for a metaheuristic algorithm provided by the population initialization, a common issue is presented in several initialization methodologies, which is commonly presented in high-dimensional optimization problems: the candidate solutions tend to cluster in determined regions, generating poor exploration patterns and premature convergence in sub-optimal regions, avoiding the finding of the global solution.

Several research works, such as the one presented by Kazimipour [16] have analyzed the performance obtained by metaheuristic algorithms employing different population initialization methods, analyzing a decreasing performance for metaheuristic algorithms when facing optimization problems with a greater number of dimensions in extended spaces presenting a poor performance as the dimensions increase. The expansion of the optimization landscape presents diverse challenges for the population, such as the poor capability of the population to explore unvisited regions in space or exploiting determined locations prematurely. This problem is presented for both deterministic and stochastic initialization techniques; given the random nature of their operators in generating the decision variables, most initialization techniques tend to generate samples for each search agent with a high degree of correlation in their decision variables.

Another issue presented in these methodologies is the grouping phenomena presented in the candidate solutions, generating poor exploration patterns and stagnation in sub-optimal regions. This problem presents a greater challenge for metaheuristic algorithms to cover adequately the solutions space.

3.5 Gibbs Sampling

Gibbs Sampling algorithm [17] is a statistical sample generation methodology belonging to the family of Monte Carlo Markov Chains (MCMC) methodologies, which generates samples from a multivariate probability distribution when it is difficult to calculate the density directly or the sampling process becomes a difficult challenge, originally proposed by Stuart and Donald German [28] giving its name from Josiah Willard Gibbs who generated a sampling process commonly employed by the statistics and physics community.

In the field of statistics, Gibbs sampling is a statistical technique for the sampling of a multivariate probability distribution when direct sampling becomes difficult because of the unknown density of the function. This method eases the sampling process by generating a conditional proposed distribution, generating a sequential sampling process considering the set of previously generated samples. In Fig. 3.6, an example of the sampling process is represented by a Gaussian normal distribution. The samples are generated in two dimensions, considering the values of the previous

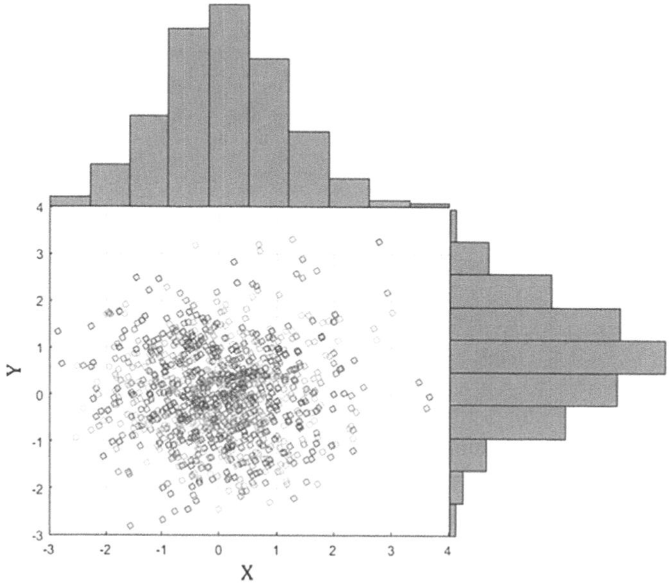

Fig. 3.6 Graphical sampling process generated by the Gibbs algorithm

sample to generate the following value and approaching the values to the proposed normalized distribution.

Gibbs Sampling method is essentially a special case of the Metropolis-Hasting methodology. Still, in a more extended perspective, it can be considered as a different sampling scheme from a set with a major number of variables, sequentially making samples, making it able to incorporate the Metropolis–Hastings acceptance criterion in the sampling process.

This univariate sampling process reflects an approximation for the generation of multiple variables in relation to the other variables. This process is looped through all the required variables, aiming to obtain a sampling sequence approximate to a proposed target distribution by conditional probability distributions, defined in Eq. (3.11):

$$\pi_i(\theta_i) = \pi\left(\theta_i|\theta_j, j \neq i\right) \tag{3.11}$$

While the traditional value generation schemes are computationally costly or difficult to implement, Gibbs sampling is an easy alternative to generate values sequentially from conditional distributions. Algorithm 1 exposes the general overview of the Gibbs Sampling framework.

Algorithm 1. Gibbs sampling methodology	
1.	Configure $t = 1, M, d$
2.	Generate an initial sample $x_1\{x_{1,1}, \ldots, x_{1,d}\}$
3.	**Repeat until** $t = M$
4.	$t = t + 1$
5.	**for** each variable $i = 1, \ldots, d$
6.	Obtain new value for $p_{t,1} = \mathbf{x}_{t-1,i-1}$
7.	**end**
8.	**end**

The Gibbs sampling algorithm is a highly effective technique for high-dimensional spaces. One of its main advantages is the ability to decompose the sampling process of a multidimensional joint distribution in a sequence of unidimensional problems. This method jointly employs the dependence among variables in a more efficient way.

3.5.1 Mathematical Example

For a better understanding of the Gibbs sampling, a mathematical representation for approaching a bivariate (x_1, x_2) Gaussian density distribution $P = N(\mu, \theta)$ is represented. For this example, a $\mu = [0{,}0]$ and $\Sigma = \begin{bmatrix} 1 & \sigma_{1,2} \\ \sigma_{2,1} & 1 \end{bmatrix}$ are considered, where $\sigma_{1,2}, \sigma_{2,1} = 0.8.$. To determine all the samples., two different probability functions are utilized to denote the relationship between the samples, described in Eq. (3.12):

$$p_{1,i} = N\left(\mu_1 + \sigma_{2,1}(x_{2,i-1} - \mu_2), \sqrt{1 - \sigma_{2,1}^2}\right)$$

$$p_{2,i} = N\left(\mu_2 + \sigma_{1,2}(x_{1,i-1} - \mu_1), \sqrt{1 - \sigma_{1,2}^2}\right) \tag{3.12}$$

In Fig. 3.7, a graphical representation of the samples generated by this method is illustrated for different samples. In (a), the sampling performed for a total of 10 samples can be observed. The samples are distributed along the space, highlighting a low correlation among themselves and preventing a premature grouping in a particular region. In (b), a 2-dimensional histogram of 100 samples is observed. The histogram presents an effective dispersion along the space of the function, highlighting the ability of the method to distribute uniformly the samples even in the early stages. As the number of samples increases, the process consequently approaches

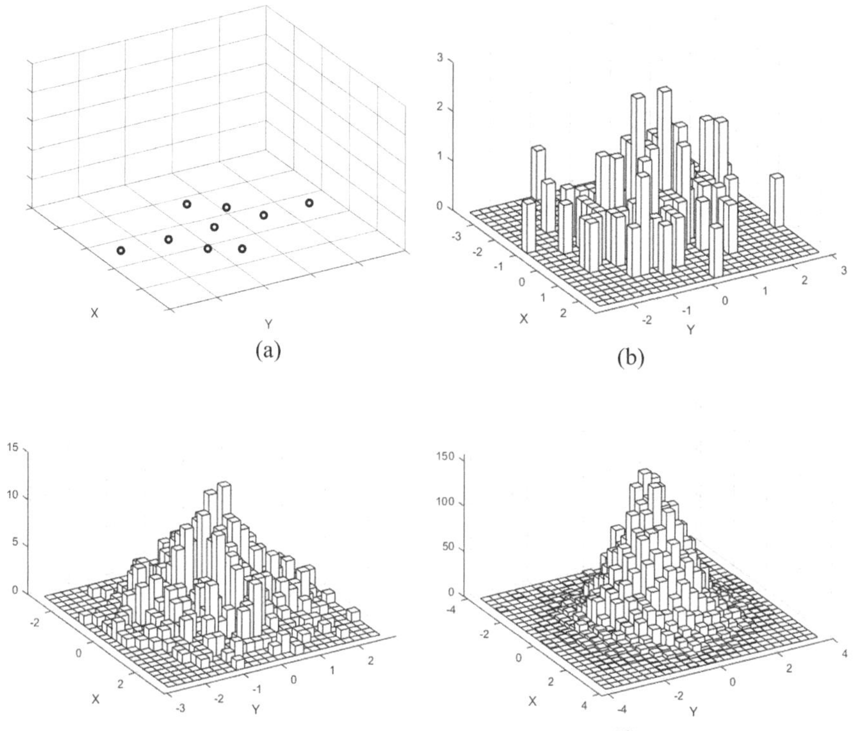

Fig. 3.7 Visual representation of the sampling process generated by the Gibbs sampling method for the distribution proposed in Eq. (3.12) for **a** 10, **b** 100, **c** 1,000 and **d** 10,000 samples

the target distribution. The convergence to the proposed distribution is evidenced in (c) and (d), where sampling histograms for 1,000 and 10,000 samples, respectively, can be observed. The histograms denote progressive refinement to the target distribution, proving how the progressive increase of samples improves the accuracy and the representation of the distribution through the Gibbs Sampling.

3.6 Gibbs Initialization Methodology

Gibbs sampling is a statistical methodology used for generating values with a low degree of correlation among the generated samples and approximating the samples to a target distribution. The low correlation provided by this method affords a better dispersion on the generated solutions, which, implemented in a metaheuristic methodology, brings more comprehensive coverage of the optimization landscape, potentially increasing the performance of the algorithm from the initial step of the evolution process. The proposed initialization scheme presents a method for generating a set of d-dimensional variables $\{x_{i,1}, \ldots, x_{i,d}\}$ for a group of n candidate positions $\{\mathbf{x}_1, \ldots, \mathbf{x}_n\}\left(\mathbf{x}_{i\in n} = \left\{x_{i,1}, \ldots, x_{i,d}\right\}\right)$ Employing the Gibbs sampler algorithm. The main target of this approach is to ensure a better knowledge of the space through a proposed Gaussian distribution covering most possible areas of the space, defined by the cost function $J(\mathbf{x}_i)$. Through the application of the Gibbs sampler to a Gaussian distribution, the initialization scheme generates the positions $\{\mathbf{x}_1, \ldots, \mathbf{x}_n\}$ obtaining a better distribution of the search space, preventing the concentration of certain individuals in a determined region.

3.6.1 Mean and Standard Deviation Calculations

The first step of the initialization method is to determine the mathematical factors involved in the individual generation. First, it is important to define the lower $\boldsymbol{P}_{i,lb}$ and upper $\boldsymbol{P}_{i,ub}$ boundaries of the search space for the optimization problem. The generation of variables is considered as a mean value of a vector $\boldsymbol{\mu}[\mu_1, \mu_2, \ldots, \mu_d]$. The values for each decision variable $\mu_i \in \boldsymbol{\mu}$ is calculated as follows in Eq. (3.13):

$$\mu_i = \frac{\boldsymbol{P}_{i,ub} - \boldsymbol{P}_{i,lb}}{2} \tag{3.13}$$

Afterwards, it is necessary to determine the values for the standard deviation por each decision variable of the search agents $\boldsymbol{\sigma} = [\sigma_1, \ldots, \sigma_d]$. The standard deviation value for each decision variable σ_i is calculated by Eq. (3.14):

$$\sigma_i = \left|\frac{P_{i,u_b} - P_{i,l_b}}{3}\right| \tag{3.14}$$

This delimitation is based on the properties of the Gaussian distribution, where approximately 99.73% of the elements are placed in a ratio of three times the standard deviation with respect to the mean ($\pm 3\sigma_i$). The reasoning for this configuration for the standard deviation to a third part of the total range is to ensure that the samples are dispersed across the breadth of the range of variables.

3.6.2 Probability Function

The probability function $p_{t,i}$ that relates the value assumed by the t variable of the current i sample with respect to the previous variable $t-1$ and the previous sample $i-1$ is defined and proposed as:

$$p_{t,i} = N\left(\frac{\mu_1 + \sigma_i}{\sigma_{i-1} - x_{t-1,i-1}}, \sqrt{\frac{1-\sigma_i}{\sigma_{i-1}}}\right) \tag{3.15}$$

The probability function established in Eq. (3.15) has the main objective of obtaining a unique output: the generation of positions not correlated among themselves by a semi-circular pattern. This pattern is achieved by adding the properties of the Gaussian distribution, such as the symmetry around all the variables and the characteristic bell shape, allowing the samples to align inside a semi-circular array.

3.6.3 Computational Procedure

Having detailed all the generalities of the methodology to generate a diverse set of initial positions, it remains to define concisely the complete computational process for the initialization. This computational description is detailed in a pseudocode exposed in Algorithm 2. The first step in Line 1 is to define the size of the search space $\mathbf{X} = \left\{\mathbf{x} \in \mathbb{R}^d | P_{j,l_b} \leq x_j \leq P_{j,u_b}\right\}$ of the cost function $J(x)$. Lines 2 and 3 establish the required parameters for the sampling process, which are the mean $\boldsymbol{\mu}$ and standard deviation $\boldsymbol{\sigma}$ vectors required for the sampling of all the decision variables forming an individual of the population. The next step is to define the first individual in the population, who may take any position in space, be the center point of space, or be in any random position. After getting this initial sample, the loop process is started for the generation of the $n-1$ remaining individuals. This process is performed through the Gibbs sampler. In Line 9, the samples are generated following the proposed Gaussian probability distribution discussed in Eq. (3.7).

An important feature of the initialization algorithm described in Algorithm 2 is that it is an efficient and simple computational process. This process does not require a great number of calculations or evaluations, which represents a great advantage by not increasing in a considerable way the complexity and computational cost required for a metaheuristic algorithm. This aspect is especially important in algorithms with high computational demand, where computational efficiency plays a crucial role in the correct performance of the algorithm, even for high-dimensional problems, without affecting the precision of the sampling method.

Algorithm 2. Gibbs initialization methodology

1.	Define $\mathbf{X}, n, d$
2.	Calculate $\boldsymbol{\mu} = [\mu_1, \ldots, \mu_d]$ by Equation (12)
3.	Calculate $\boldsymbol{\sigma} = [\sigma_1, \ldots, \sigma_d]$ by Equation (13)
4.	Configure $t = 1$
5.	Generate an initial sample for $\mathbf{x_1}$
6.	**for** $t = n$
7.	$t = t + 1$
8.	**for** $i = 1, \ldots, d$
9.	$x_{t,i} = p_{t,i} = (\mathbf{x_{t-1,i-1}})$ (Equation (14))
10.	**End**
11.	**end**

3.6.4 Implementation in a Metaheuristic

Once the initialization process has created a full population of d-dimensional n individuals, it is required to validate the correct location of the search agents to find an optimal solution more efficiently. For an easy validation, this initialization method has been embedded in a traditional Differential Evolution (DE) algorithm [29]. The selection for this algorithm is given to the simplicity in the operators and efficiency to explore most areas of the search space and identify easily the promising regions for most complex optimization problems. This algorithm has been chosen to prove the performance of this initialization technique easily.

The mutation scheme employed for the evaluation is the $DE/best/1$. This mutation scheme was isolated with the objective of utilizing a consistent algorithmic base throughout different methods, obtaining a more objective evaluation. The parameters set for the evaluation are described in Table 3.1. The evaluation of this method will employ a population of 50 individuals, comparing two different dimensionalities, these 30 and 100 dimensions. The reason for employing these numbers is to evaluate the efficiency and how an initialization method hay improve or affect the performance of a metaheuristic algorithm even when a high number of decision variables are employed for the individuals. The F escalation factor has been set with a significantly low value to present a great dispersion of the mutant vectors, avoiding the new solutions passing by the optimal regions without evaluating them.

Table 3.1 Initial parameters employed for the DE algorithm

Mutation factor	$DE/Best/1$
Population size n	50
Dimensionality d	30,100
Cross-over probability pCr	0.9
Escalation factor F	0.2

3.7 Experimental Results

This section is intended to summarize the results obtained from the experimentation process applied to a metaheuristic algorithm for solving complex optimization problems. For the sake of a fair comparison, the experimentation will compare the DE algorithm with populations generated by following different methodologies, using the parameters indicated in Table 3.1 for all the scenarios. The performance comparative of the proposed methodology against a group of different techniques is performed to demonstrate the relative strengths and potential advantages when the algorithm faces complex scenarios.

The initialization method named GIBBS has been applied to the Differential Evolution (DE), and extensive validation has been performed using a set of diverse initialization methodologies applied to the same DE algorithm. The comparison has been performed competing against the most common random initialization (RI) described in Eq. (3.1), which is the most commonly employed initialization in most novel metaheuristic algorithms, the Opposition-Based (OP) initialization, the Adaptive Randomness (AR) initialization, the Chaotic (CM) methodology and the Metropolis–Hastings (MH) initialization approach, which are well-established initialization approaches from the literature, proving all of them significant improvements in the performance and the quality of the solutions.

3.7.1 *Comparison in Terms of Quality of Solutions*

To perform a comparative analysis of the performance of the methodologies, they have been tested employing a set of 30 objective functions extracted from the traditional benchmarking test set [30] selected by the combination of different landscapes with different degrees of complexity, including multimodal, unimodal and hybrid functions. The set of functions has been labeled from $f(x)_1$ to $f(x)_{30}$ and detailed in Table 3.8 of the Appendix. This set of functions includes 23 multimodal, four unimodal, and three hybrid functions, each representing a unique challenge for evaluating the performance of the methodology.

It is worth noting that the multimodal functions represent the greatest challenge for the optimization methodologies, presenting multiple regions with promising solutions but only one global solution, increasing the susceptibility to getting stuck in a local optimal solution. In the analysis, the correct configuration and performance of the different initialization techniques are tested in terms of the accuracy of the obtained solutions upon these functions, focusing especially on the ability of each method to place strategically the solutions facilitating the process to obtain the global optimal solution from all the local optimal solutions.

The experimentation has been performed considering 30 independent executions of these initialization approaches, generating a new population for all the methods, aiming to test the robustness of the algorithms employing a different initialization, preventing a favorable result in a single execution obtained by the stochastic factors of the initialization, ensuring that the proposed initialization affords constantly good-quality solutions for the metaheuristic algorithm. For all the algorithms, a stop criterion has been set to 50,000 function accesses.

Since each algorithm is tested 30 independent times for all the functions, the analysis considers three different factors presenting different characteristics of the performance of each algorithm: The Average Best-So-Far (*AB*), the Median (*MD*) and the Standard Deviation (*SD*). The *AB* and *MD* are employed to assess the accuracy of the best solutions obtained against the expected ones. The *AB* is the average of the solutions obtained for all the 30 different executions. Meanwhile, the *MD* is the median of all the obtained solutions sorted. On the other hand, the *SD* reflects the dispersion existing in the obtained best solutions, determining, by consequence, the robustness of each algorithm. This multi-phase exhaustive analysis is performed for all the optimization methods to better comprehend the efficiency of each initialization in different optimization landscapes.

As can be seen by the results obtained in Table 3.2, the DE-GIBBS methodology presents a notorious improvement for most of the study cases. The GIBBS initialization methodology presents a strongly competitive performance compared to the MH methodology, which presents a strong performance by the combination of probabilistic selection based on the solutions. However, the GIBBS initialization presents a better ability to place the solutions in strategic positions, facilitating the finding of feasible regions in a minor time that requires a minor number of evaluations.

The outstanding performance of the GIBBS methodology can be attributed to its ability to generate sets of candidate solutions preventing the possible grouping of individuals in determined zones, which is different from other initialization methodologies that often present this issue. Instead, the sampling process uniformly distributes the solutions following the Gaussian bell distribution through all dimensions. This deployment is important, especially in high-dimensional problems, likewise allowing a better exploration of space. In perspective, other initialization methods, despite their notorious location and outstanding performance in low-dimensional spaces, often tend to present a decrease in the location of the candidate solutions, provoking a lower performance for the algorithm in high-dimensional spaces; being this caused by their tendency to concentrate certain numbers of individuals in determined regions.

Table 3.2 Optimization results for a set of 30 functions with $d = 30$

Functions		DE_RI	DE_AR	DE_CM	DE_OP	DE_MH	DE_GIBBS
$f(x)_1$	AB	3.104E-02	3.104E-02	3.905E-08	3.968E-08	6.438E-09	**2.154E-09**
	MD	3.019E-08	3.852E-08	3.125E-08	2.898E-08	5.389E-09	**1.596E-09**
	SD	1.700E-01	1.700E-01	3.369E-08	2.878E-08	5.939E-09	**1.700E-09**
$f(x)_2$	AB	1.028E-01	6.758E-01	7.078E-01	8.205E-01	7.025E-01	**6.752E-01**
	MD	6.666E-01	6.666E-01	6.666E-01	6.666E-01	6.666E-01	**6.666E-01**
	SD	1.296E-01	4.972E-02	1.653E-01	5.962E+08	1.921E-01	**4.712E-02**
$f(x)_3$	AB	2.217E-03	2.054E-03	1.395E-03	1.150E-03	1.313E-03	**0.000E+00**
	MD	5.817E-14	4.379E-14	4.607E-14	5.135E-14	**0.000E+00**	**0.000E+00**
	SD	4.308E-03	3.888E-03	4.556E-03	3.086E-03	4.514E-03	**0.000E+00**
$f(x)_4$	AB	1.064E-08	8.680E-12	7.685E-09	4.079E-08	**0.000E+00**	**0.000E+00**
	MD	3.814E-15	1.317E-14	5.765E-15	7.551E-18	**0.000E+00**	**0.000E+00**
	SD	4.426E-08	3.666E-11	3.905E-08	3.9682E-08	**0.000E+00**	**0.000E+00**
$f(x)_5$	AB	9.959E-02	7.549E-02	1.603E-01	1.025E-01	9.340E-02	**8.952E-03**
	MD	4.463E-15	2.076E-15	1.202E-15	2.365E-15	3.543E-16	**9.273E-18**
	SD	1.820E-01	1.635E-01	4.420E-01	1.841E-01	3.467E-01	**2.731E-02**
$f(x)_6$	AB	**2.000E+00**	**2.000E+00**	**2.000E+00**	**2.000E+00**	**2.000E+00**	**2.000E+00**
	MD	**2.000E+00**	**2.000E+00**	**2.000E+00**	**2.000E+00**	**2.000E+00**	**2.000E+00**
	SD	**0.000E+00**	**0.000E+00**	**0.000E+00**	**0.000E+00**	**0.000E+00**	**0.000E+00**
$f(x)_7$	AB	**2.000E+00**	**2.000E+00**	**2.000E+00**	**2.000E+00**	**2.000E+00**	**2.000E+00**
	MD	**2.000E+00**	**2.000E+00**	**2.000E+00**	**2.000E+00**	**2.000E+00**	**2.000E+00**

(continued)

Table 3.2 (continued)

Functions		DE_RI	DE_AR	DE_CM	DE_OP	DE_MH	DE_GIBBS
	SD	**0.000E+00**	**0.000E+00**	**0.000E+00**	**0.000E+00**	**0.000E+00**	**0.000E+00**
$f(x)_8$	AB	**0.000E+00**	**0.000E+00**	1.440E-132	**0.000E+00**	1.578E-125	1.429E-140
	MD	**0.000E+00**	**0.000E+00**	**0.000E+00**	**0.000E+00**	2.718E-149	1.165E-149
	SD	**0.000E+00**	**0.000E+00**	7.931E-132	**0.000E+00**	8.646E-125	7.745E-140
$f(x)_9$	AB	3.321E+03	8.420E+01	7.362E+01	7.857E+01	7.223E+01	**7.103E+01**
	MD	7.368E+01	7.369E+01	7.368E+01	7.368E+01	7.236E+01	**7.103E+01**
	SD	1.778E+04	3.679E+01	1.975E+00	1.894E+01	1.214E+00	**3.296E-06**
$f(x)_{10}$	AB	1.076E+02	1.064E+02	1.056E+02	1.061E+02	**1.055E+02**	1.096E+02
	MD	1.055E+02	1.073E+02	1.060E+02	1.066E+02	**1.059E+02**	1.090E+02
	SD	1.543E+01	3.255E+00	2.804E+00	4.589E+00	3.759E+00	**2.176E+00**
$f(x)_{11}$	AB	8.070E+46	1.747E+50	3.141E+54	2.321E+47	2.887E+17	**4.565E+01**
	MD	1.976E+31	8.361E+28	2.485E+35	1.827E+27	1.161E+03	**3.003E+01**
	SD	4.383E+47	9.571E+50	1.720E+55	1.271E+48	1.578E+18	**7.395E+01**
$f(x)_{12}$	AB	**3.000E+01**	**3.000E+01**	**3.000E+01**	**3.000E+01**	**3.000E+01**	**3.000E+01**
	MD	**3.000E+01**	**3.000E+01**	**3.000E+01**	**3.000E+01**	**3.000E+01**	**3.000E+01**
	SD	**0.000E+00**	**0.000E+00**	**0.000E+00**	**0.000E+00**	**0.000E+00**	**0.000E+00**
$f(x)_{13}$	AB	8.027E-04	6.942E-04	8.591E-04	8.408E-04	7.346E-04	**3.420E-04**
	MD	5.936E-04	6.562E-04	7.088E-04	7.970E-04	6.273E-04	**3.342E-04**

(continued)

Table 3.2 (continued)

Functions		DE_RI	DE_AR	DE_CM	DE_OP	DE_MH	DE_GIBBS
	SD	5.306E-04	3.193E-04	5.491E-04	4.430E-04	4.822E-04	**1.385E-04**
$f(x)_{14}$	AB	6.374E-02	1.363E-01	2.101E-01	1.463E-02	**3.032E-07**	6.177E+00
	MD	6.077E-06	1.121E-05	4.705E-07	2.189E-05	**3.286E-10**	1.005E-07
	SD	3.461E-01	7.054E-01	1.145E+00	6.344E-02	**9.440E-07**	2.331E+01
$f(x)_{15}$	AB	9.861E+00	9.878E+00	9.702E+00	9.771E+00	9.836E+00	**9.719E+00**
	MD	9.912E+00	9.932E+00	9.771E+00	9.787E+00	9.842E+00	**9.722E+00**
	SD	3.954E-01	4.983E-01	4.421E-01	3.964E-01	**3.453E-01**	4.4779E-01
$f(x)_{16}$	AB	2.055E-04	1.781E-04	9.478E-05	6.560E-05	1.741E-04	**4.684E-06**
	MD	5.845E-05	4.197E-05	2.209E-05	2.552E-05	1.331E-05	**1.667E-06**
	SD	3.882E-04	6.696E-04	1.468E-04	9.265E-05	6.184E-04	**8.562E-06**
$f(x)_{17}$	AB	1.513E+02	1.456E+02	1.521E+02	1.483E+02	1.488E+02	**1.220E+02**
	MD	1.542E+02	1.508E+02	1.636E+02	1.491E+02	1.485E+02	**1.263E+02**
	SD	2.777E+01	2.823E+01	2.999E+01	2.092E+01	2.433E+01	2.781E+01
$f(x)_{18}$	AB	3.067E+01	3.023E+01	3.253E+01	2.946E+01	2.939E+01	**2.557E+01**
	MD	2.589E+01	2.505E+01	2.596E+01	**2.490E+01**	2.595E+01	2.552E+01
	SD	1.546E+01	1.699E+01	1.791E+01	1.666E+01	1.188E+01	**1.005E+00**
$f(x)_{19}$	AB	1.174E+01	1.146E+01	1.087E+01	1.107E+01	9.840E-01	**1.973E-01**
	MD	1.075E+01	1.048E+01	9.730E+00	1.123E+01	9.406E-01	**1.728E-01**
	SD	3.954E+00	5.457E+00	5.156E+00	4.854E+00	5.024E-01	**7.606E-02**
$f(x)_{20}$	AB	2.592E-05	2.176E-05	3.167E-05	3.325E-05	9.795E-09	**3.684E-09**

(continued)

Table 3.2 (continued)

Functions		DE_RI	DE_AR	DE_CM	DE_OP	DE_MH	DE_GIBBS
	MD	1.132E-05	4.472E-06	1.152E-05	1.028E-05	7.967E-09	**3.249E-09**
	SD	3.743E-05	5.598E-05	4.115E-05	6.446E-05	5.210E-09	**2.247E-09**
$f(x)_{21}$	AB	1.000E-01	1.333E-01	1.666E-01	2.000E-01	3.666E-01	**0.000E+00**
	MD	0.000E+00	0.000E+00	0.000E+00	0.000E+00	0.000E+00	**0.000E+00**
	SD	3.051E-01	3.457E-01	3.790E-01	7.611E-01	9.643E-01	**0.000E+00**
$f(x)_{22}$	AB	−1.144E+03	−1.145E+03	−1.150E+03	−1.145E+03	−1.075E+03	**-1.153E+03**
	MD	−1.146E+03	−1.146E+03	−1.160E+03	−1.146E+03	−1.061E+03	**-1.153E+03**
	SD	2.188E+01	2.472E+01	2.316E+01	2.130E+01	9.8409E-01	**1.9734E-01**
$f(x)_{23}$	AB	9.517E-01	1.158E+00	7.6509E-01	1.239E+00	2.905E-01	**8.683E-03**
	MD	6.987E-01	8.770E-01	6.0753E-01	1.064E+00	2.017E-01	**5.950E-03**
	SD	7.097E-01	8.265E-01	6.6236E-01	8.307E-01	2.578E-01	**7.829E-03**
$f(x)_{24}$	AB	1.966E-12	1.534E-13	1.4091E-13	1.353E-13	9.218E-16	**1.152E-16**
	MD	7.279E-14	5.217E-14	4.6457E-14	5.063E-14	5.219E-16	**5.592E-17**
	SD	9.826E-12	2.656E-13	2.5094E-13	1.866E-13	1.102E-15	**2.292E-16**
$f(x)_{25}$	AB	7.442E-12	4.386E-12	7.8825E-12	4.083E-12	6.505E-15	**1.609E-15**
	MD	2.233E-12	1.942E-12	3.3320E-12	2.285E-12	3.860E-15	**7.793E-16**
	SD	1.857E-11	7.213E-12	1.2563E-11	6.549E-12	5.482E-15	**1.627E-15**
$f(x)_{26}$	AB	5.4865E-17	6.830E-17	1.1697E-16	6.184E-17	5.063E-17	**7.368E-18**
	MD	3.359E-17	3.119E-17	4.0023E-17	4.057E-17	3.674E-17	**5.121E-18**
	SD	8.663E-17	8.718E-17	3.1370E-16	6.624E-17	4.781E-17	**7.670E-18**

(continued)

Table 3.2 (continued)

Functions		DE_RI	DE_AR	DE_CM	DE_OP	DE_MH	DE_GIBBS
$f(x)_{27}$	AB	1.330E-14	7.612E-15	2.7396E-15	2.279E-15	1.269E-15	**1.072E-16**
	MD	1.474E-15	1.615E-15	2.0853E-15	1.690E-15	5.622E-16	**3.162E-17**
	SD	5.747E-14	2.655E-14	2.7001E-15	2.110E-15	3.396E-15	**2.348E-16**
$f(x)_{28}$	AB	4.164E-08	3.328E-08	3.8870E-08	4.547E-08	8.824E-10	**3.599E-10**
	MD	3.345E-08	2.777E-08	3.3243E-08	3.513E-08	7.918E-10	**3.449E-10**
	SD	2.988E-08	2.565E-08	1.7881E-08	2.554E-08	4.615E-10	**1.769E-10**
$f(x)_{29}$	AB	3.490E+01	3.415E+01	3.385E+01	3.439E+01	**3.200E+01**	**3.200E+01**
	MD	3.321E+01	3.337E+01	3.340E+01	3.331E+01	**3.200E+01**	**3.200E+01**
	SD	6.484E+00	3.041E+00	1.689E+00	2.782E+00	4.761E-03	**2.730E-07**
$f(x)_{30}$	AB	3.439E+01	3.422E+01	3.309E+01	3.529E+01	**2.900E+01**	**2.900E+01**
	MD	2.900E+01	2.900E+01	2.900E+01	2.900E+01	**2.900E+01**	**2.900E+01**
	SD	8.866E+00	6.988E+00	7.198E+00	8.205E+00	8.724E-10	**3.940E-10**

3.7.2 Comparison in Terms of Dimensionality

This section summarizes an analysis of the results obtained by the algorithms employing different population initialization techniques for high-dimensional optimization problems.

This assessment is performed employing a set of 14 different test functions detailed in Table 3.8 of the Appendix, combining eight multimodal, three unimodal and two hybrid test functions, labeled from $f(x)_1$ to $f(x)_9$, and from $f(x)_{24}$ and $f(x)_{29}$. This dimensionality analysis is intended to demonstrate the advantages of a correct initial configuration from the population initialization for the metaheuristic algorithms to reach high-quality solutions even in more complex spaces.

For this comparison, three fundamental factors are involved for a better analysis of the performance. The Average Best-So-Far (*AB*), the Median (*MD*) and the Standard Deviation (*SD*). This comparison has been performed by testing the initialization methods in 100-dimensional optimization spaces. Table 3.3 summarizes the performance obtained for the initialization methods employing diverse cost functions.

By the results observed in Table 3.3, it can be observed that there is a notorious improvement in the GIBBS methodology for 12 of the proposed study cases. This initialization methodology outstands in high-dimensional spaces where other initialization techniques begin to present certain difficulties for reaching optimal solutions, getting trapped in suboptimal solutions. It is worth noting a comparable performance to the MH methodology; both MH and GIBBS sharing certain similarities with the MCMC sampling methods. However, GIBBS presents a better distribution of the solutions, enhancing the performance of the metaheuristic algorithm.

The low degree of correlation on the sampling process obtained by the GIBBS approach affords the algorithm the capability to create by simple calculations a set of candidate solutions with an adequate distribution along the space, placing the solutions strategically in a space, obtaining a wider spectrum in the information of the space by the obtained initial solutions. When facing high-dimensional spaces, the optimization process becomes a very challenging task for obtaining solutions with promising quality, and the initial configuration plays a fundamental role in the design of strong evolutionary mechanisms capable of locating efficiently promising solutions.

3.7.3 Evaluation in Terms of Convergence Rapidness and Robustness

This section examines the performance of the initialization methodologies from another focus, considering the convergence patterns of the population to an optimal position. This includes an evaluation of the required cost function evaluations until

Table 3.3 Optimization results for a set of 15 functions with $d = 100$

Functions		DE_RI	DE_AR	DE_CM	DE_OP	DE_MH	DE_GIBBS
$f(x)_1$	AB	2.848E+00	3.287E+00	2.974E+00	2.777E+00	2.631E+00	**1.494E+00**
	MD	2.896E+00	3.251E+00	2.970E+00	2.697E+00	2.402E+00	**1.727E+00**
	SD	7.250E-01	8.596E-01	7.763E-01	7.388E-01	1.268E+00	**8.080E-01**
$f(x)_2$	AB	2.973E+02	3.456E+02	2.820E+02	2.629E+02	1.126E+02	**2.027E+01**
	MD	2.202E+02	2.535E+02	2.102E+02	1.692E+02	8.537E+01	**1.994E+01**
	SD	2.086E+02	4.103E+02	2.163E+02	2.764E+02	1.156E+02	**1.144E+01**
$f(x)_3$	AB	9.994E-01	1.188E+00	1.039E+00	1.036E+00	1.047E+00	**2.274E-01**
	MD	1.069E+00	1.070E+00	1.026E+00	1.016E+00	1.047E+00	**1.690E-01**
	SD	2.810E-01	6.250E-01	2.727E-01	2.342E-01	1.025E-02	**2.190E-01**
$f(x)_4$	AB	3.916E-06	2.730E-05	4.394E-05	2.692E-05	**0.000E+00**	**0.000E+00**
	MD	0.000E+00	0.000E+00	0.000E+00	0.000E+00	**0.000E+00**	**0.000E+00**
	SD	1.943E-05	5.768E-05	1.472E-04	7.722E-05	**0.000E+00**	**0.000E+00**
$f(x)_5$	AB	9.249E+00	8.363E+00	1.138E+01	8.089E+00	5.476E+00	**1.148E+00**
	MD	8.640E+00	8.149E+00	1.174E+01	7.792E+00	5.355E+00	**1.211E+00**
	SD	3.030E+00	3.290E+00	3.311E+00	2.727E+00	2.415E+00	**5.965E-01**
$f(x)_6$	AB	2.000E+00	2.006E+00	2.000E+00	2.003E+00	8.131E+33	**2.000E+00**
	MD	2.000E+00	2.002E+00	2.000E+00	2.002E+00	5.315E+22	**2.000E+00**
	SD	9.501E-05	1.947E-03	1.667E-04	3.945E-04	4.319E+34	**8.926E-05**
$f(x)_7$	AB	2.003E+00	2.002E+00	2.003E+00	2.002E+00	2.587E+40	**2.002E+00**
	MD	2.002E+00	2.002E+00	2.002E+00	2.002E+00	4.866E+19	**2.002E+00**
	SD	1.721E-04	1.031E-04	1.460E-04	1.744E-04	1.417E+41	**8.657E-05**
$f(x)_8$	AB	**0.000E+00**	**0.000E+00**	0.000E+00	**0.000E+00**	1.545E-219	2.106E-260
	MD	**0.000E+00**	**0.000E+00**	0.000E+00	**0.000E+00**	1.877E-247	6.367E-278
	SD	**0.000E+00**	**0.000E+00**	0.000E+00	**0.000E+00**	0.000E+00	0.000E+00
$f(x)_9$	AB	2.964E+05	1.020E+06	**2.574E+05**	4.063E+05	4.288E+06	4.219E+05
	MD	1.345E+05	1.735E+05	**8.269E+04**	1.898E+05	1.170E+05	9.582E+03
	SD	4.012E+05	2.367E+06	**4.168E+05**	5.902E+05	2.185E+07	1.723E+06
$f(x)_{24}$	AB	2.886E+02	3.818E+02	2.386E+02	4.079E-08	9.556E+03	**6.439E+00**
	MD	1.697E+02	1.907E+02	1.811E+02	7.551E-18	9.475E+03	**1.296E+00**
	SD	3.256E+02	6.462E+02	1.792E+02	3.9682E-08	2.214E+03	**1.335E+01**
$f(x)_{25}$	AB	2.087E+04	3.828E+04	3.240E+04	1.025E-01	6.773E+05	**2.902E+03**
	MD	1.052E+04	1.193E+04	1.238E+04	2.365E-15	6.852E+05	**3.212E+02**
	SD	3.112E+04	6.880E+04	4.727E+04	1.841E-01	1.504E+05	**6.352E+03**
$f(x)_{26}$	AB	4.423E-02	5.140E-02	5.404E-02	**2.000E+00**	4.512E-02	**7.141E-03**
	MD	1.895E-02	1.666E-02	2.036E-02	**2.000E+00**	2.227E-02	**4.166E-03**
	SD	9.230E-02	1.422E-01	1.079E-01	**0.000E+00**	6.318E-02	**1.130E-02**
$f(x)_{28}$	AB	5.077E+01	4.133E+01	1.634E+02	**2.000E+00**	3.133E+02	**1.180E+00**

(continued)

Table 3.3 (continued)

Functions		DE_RI	DE_AR	DE_CM	DE_OP	DE_MH	DE_GIBBS
	MD	3.407E+01	2.214E+01	2.463E+01	**2.000E+00**	3.107E+02	**3.331E-01**
	SD	5.713E+01	4.619E+01	3.395E+02	**0.000E+00**	4.843E+01	**2.028E+00**
$f(x)_{29}$	AB	7.542E+04	1.969E+04	4.156E+04	**0.000E+00**	6.079E+03	**1.094E+03**
	MD	1.656E+04	1.159E+04	1.625E+04	**0.000E+00**	3.997E+03	**9.372E+02**
	SD	2.608E+05	3.818E+02	2.386E+02	**0.000E+00**	9.556E+03	**6.439E+00**

an optimal solution is found and the rapidness of the members of the population to converge in a determined region.

For this evaluation, three different factors are considered to estimate the behaviors and the convergence patterns of the search agents to zones identified as promising: The number of function calls (NFC), Success Rate (SR) and the Acceleration Rate (ARE). The assessment of these aspects allows us to comprehend not only the effectiveness of the initialization methods to find optimal solutions but also the efficiency and liability of search operators. This analysis is fundamental to identifying strengths and potential areas of improvement in the initialization methods used for the optimization algorithms.

The NFC number is a counter indicating how many times the optimization algorithm is required to evaluate the cost function with an individual of the population and validate the efficiency of optimization algorithms. To obtain this metric, the process of the metaheuristic algorithms is configured with a VTR or Value-To-Reach. This VTR value is a target or expected solution obtained during the optimization process. Since all algorithms present variability in their outputs by different optimization problems, it is required to set different VTR values for the employed test functions. Table 3.7 summarizes the VTR values employed for the set of 30 test functions used in the comparison of the quality of solutions. These values were obtained based on the observed performance from each methodology, obtaining the best configuration to measure adequately the results for each test function.

An algorithm with a lower NFC value indicates better effectiveness in its evolutionary process. This implies a minor requirement for the evaluation of the cost function from the algorithm to reach an optimal solution. This effectiveness is particularly significant in complex optimization duties where the computational resources play a vital role in proper performance. The NFC not only indicates a faster convergence but also reflects the ability to explore the searching space efficiently, aiming to reach the optimal locations with a minor computational demand. Aiming to keep a liable and accurate evaluation of the performance from all the algorithms, the average of the NFC values for a set of 30 independent executions is calculated with the objective of reducing favorable scenarios caused by the stochastic nature of the initialization methods.

The Success Rate (SR) is another widely used parameter to estimate the effectiveness of an algorithm quantitatively. In general, this parameter indicates the percentage of executions that an algorithm successfully surpasses the *VTR* value. A high *SR* represents a strong indicator of the liability and consistency of an algorithm to converge in the global optimal solution. The *SR* value is standardized and is calculated as follows:

$$SR = \frac{NSF}{TE} \tag{3.16}$$

where *NSF* corresponds to the number of successful executions where the *VTR* threshold is surpassed, and *TE* corresponds to the total number of executions per algorithm.

The third factor employed for the evaluation of the convergence pattern for the optimization algorithms is the Acceleration Rate (*ARE*). This metric is designed with the objective of comparing the efficiency of two different methodologies, denoted as *A* and *B*, based on their respective *NFC* values. The *ARE* factor is determined as follows:

$$ARE_{A/B} = \frac{NFC_A}{NFC_B} \tag{3.17}$$

where NFC_A and NFC_B correspond to the *NFC* values reported by both methodologies *A* and *B*. A $ARE_{A/B}$ upper than 1 indicates a better performance obtained by the method *B* compared to the method *A*.

Table 3.5 summarizes the *NFC* and *SR* values obtained by the algorithms employing the analyzed initialization methodologies for an extract of 26 functions detailed in Table 3.8 of the Appendix for a set of 30 independent executions from the analysis performed in Sect. 3.7.1 considering the *VTR* values established in the Table 3.4.

Complementary to the information of the *NFC* and *SR* values, Table 3.6 summarizes the numerical results for the representative set of 12 functions in terms of the *ARE* factor calculated following the Eq. (3.10).

The results obtained in Tables 3.5 and 3.6 evidence a significant improvement in the performance from the GIBBS initialization in terms of the minor *NFC* number

Table 3.4 Initial parameters employed for the DE algorithm

Functions	*VTR* value
$f(x)_{1,3,4,5,8,14,16,21,24,25,26,27,28}$	10^{-6}
$f(x)_{2,23}$	0.7
$f(x)_{6,7}$	2
$f(x)_{12,18,37}$	30
$f(x)_{13,20}$	10^{-3}
$f(x)_{22}$	−1100
$f(x)_{29}$	33
$f(x)_{15,19}$	10

Table 3.5 Numerical *NFC* and *SR* results for a representative set of 26 functions

Function	DE_RI		DE_AR		DE_CM		DE_OP		DE_MH		DE_GIBBS	
	NFC	*SR*	*NFC*	*SR*			*NFC*	*SR*	*NFC*	*SR*	*NFC*	*SR*
$f(x)_1$	41,808.33	0.9667	41,685	1	42,705.23	1	42,308.33	1	37,340	1	**34,513.33**	**1**
$f(x)_2$	18,865	0.8333	19,853.33	0.9333	21,645	0.9	22,363.93	0.9333	20,030	0.9667	**14,850**	**1**
$f(x)_3$	22,918.33	0.7667	22,858.33	0.7667	21,141.67	0.7	26,430	0.8667	16,621.66	0.9	**15,635**	**1**
$f(x)_4$	8741.66	1	8273.33	1	7473.333	1	7767.86	0.9667	200	1	**100**	**1**
$f(x)_5$	16,438.33	0.6333	18,940	0.7333	17,580	0.7564	**16,284.6**	0.6333	16,468.33	0.7	18,498.33	**0.9**
$f(x)_6$	43,653.33	1	43,698.33	1	44,390.06	1	43,695.8	1	43,418.33	1	**42,806.66**	**0.9667**
$f(x)_7$	44,498.33	1	44,635	1	45,642.8	1	44,266.33	1	44,361.66	1	**43,558.33**	**0.9667**
$f(x)_8$	288.33	1	308.33	1	432	1	428.33	1	1033.33	1	**100**	**1**
$f(x)_{12}$	6141.66	1	6166.66	1	6459.3	1	5854.53	0.9667	6330	1	**3898.33**	**1**
$f(x)_{13}$	34,586.66	0.8	32,130	0.8	36,421.67	0.8333	32,114.6	0.7333	**28,691.66**	0.7	31,960	**0.9667**
$f(x)_{14}$	14,026.66	0.3	**12,690**	0.2667	14,181.67	0.3	15,490	0.3333	35,960	**0.9**	27,416.66	0.6333
$f(x)_{15}$	15,600	0.6333	13,836.66	0.6	20,480	0.8	19,917.6	0.7667	**1351.66**	0.0333	16,041.66	0.6333
$f(x)_{16}$	–	–	–	0.0333	–	–	1668.47	–	**1261.66**	0.8666	14,300	0.3
$f(x)_{18}$	14,308.33	0.7333	16,515	0.8333	14,543.33	0.9333	12,679.73	0.7667	13,175	0.9	**10,838.33**	1
$f(x)_{19}$	5168.33	0.3667	8050	0.5333	5338.333	0.3667	4063.33	0.2667	**50**	1	**50**	1
$f(x)_{20}$	40,068.33	1	39,948.33	1	40,016.67	1	39,973.33	1	23,831.66	1	**21,408.33**	1
$f(x)_{21}$	11,601.66	0.8333	12,718.33	0.9333	11,091.67	0.8333	12,077.53	0.8667	5798.33	0.9	**3406.66**	**0.93333**
$f(x)_{22}$	17,591.66	1	14,871.66	0.9333	13,843.33	0.9667	16,655.26	1	**3486.66**	0.26667	13,548.33	**1**
$f(x)_{23}$	21,910	0.4667	**16,010**	0.3333	36,885	0.8	42,705.23	22,071.13	34,766.66	0.8333	23,026.66	**1**
$f(x)_{24}$	31,168.33	1	30,908.33	1	31,231.34	1	21,645	30,520	25,100	1	**22,508.33**	**1**
$f(x)_{25}$	35,033.33	1	34,880	1	37,653.27	1	21,141.67	34,578.33	27,693.33	1	**25,765**	**1**

(continued)

Table 3.5 (continued)

Function	DE_RI		DE_AR		DE_CM		DE_OP		DE_MH		DE_GIBBS	
	NFC	*SR*	*NFC*	*SR*			*NFC*	*SR*	*NFC*	*SR*	*NFC*	*SR*
$f(x)_{26}$	21,755	1	21,901.66	1	22,543.25	1	7473.333	21,970	21,933.33	1	**19,531.66**	**1**
$f(x)_{27}$	26,208.33	1	26,533.33	1	27,458.32	1	17,580	26,331.66	24,971.66	1	**22,263.33**	**1**
$f(x)_{28}$	42,475	1	42,280	1	42,693.33	1	44,390.06	42,761.66	34,303.33	1	**31,931.66**	**1**
$f(x)_{29}$	16,575	0.3667	16,966.66	0.36667	18,638.33	0.4	45,642.8	21,835.6	28,370	1	**15,305**	**1**
$f(x)_{30}$	13,175	0.6667	12,308.33	0.63333	13,721.67	0.7	432	11,769.93	6906.66	1	**4681.66**	**1**

Table 3.6 Numerical *ARE* results for a representative set of 12 functions

Function	GIBBS versus RI ($ARE_{RI/GIBBS}$)	GIBBS versus AR ($ARE_{AR/GIBBS}$)	GIBBS versus CM ($ARE_{CM/GIBBS}$)	GIBBS versus OP ($ARE_{OP/GIBBS}$)	GIBBS versus MH ($ARE_{MH/GIBBS}$)
$f(x)_1$	1.211367587	1.207794089	1.23735455	1.22585474	1.081900715
$f(x)_2$	1.27037037	1.336924804	1.45757576	1.50598878	1.348821549
$f(x)_3$	1.465835199	1.461997655	1.35220126	1.69043812	1.063106279
$f(x)_4$	87.41666667	82.73333333	74.7333333	77.6786667	2
$f(x)_5$	0.888638616	1.023876025	0.95035589	0.88032796	0.890260384
$f(x)_6$	1.019778851	1.020830089	1.03698941	1.02077091	1.014288974
$f(x)_7$	1.021580256	1.024717811	1.04785345	1.01625407	1.018442548
$f(x)_8$	2.883333333	3.083333333	4.32	4.28333333	10.33333333
$f(x)_{12}$	1.575459598	1.581872595	1.65693886	1.50180419	1.623770842
$f(x)_{13}$	1.082186066	1.005319149	1.13960159	1.0048373	0.897736754
$f(x)_{14}$	0.511610942	0.462857143	0.51726444	0.5649848	1.311610942
$f(x)_{15}$	0.972467532	0.862545455	1.27667532	1.24161662	0.08425974
$f(x)_{16}$	–	–	–	0.11667622	0.088227972
$f(x)_{18}$	1.320159926	1.523758265	1.34184223	1.16989697	1.215592803
$f(x)_{19}$	103.3666667	161	106.766667	81.2666667	1
$f(x)_{20}$	1.8716232	1.866017906	1.86920981	1.86718568	1.113195796
$f(x)_{21}$	3.405577299	3.733365949	3.25587084	3.54526419	1.702054795
$f(x)_{22}$	1.298437692	1.097674991	1.0217739	1.22932218	0.257350228
$f(x)_{23}$	0.951505501	0.695280834	1.60183845	0.95850318	1.50984366
$f(x)_{24}$	1.38474639	1.373195113	1.38754565	1.35594224	1.11514254
$f(x)_{25}$	1.359725726	1.3537745	1.4614116	1.34206611	1.074843133
$f(x)_{26}$	1.113832238	1.121341411	1.15418978	1.12484	1.12296271
$f(x)_{27}$	1.177197185	1.191795179	1.23334272	1.18273694	1.121649948
$f(x)_{28}$	1.330184248	1.324077457	1.33702166	1.33916175	1.074273188
$f(x)_{29}$	1.082979418	1.108570184	1.21779375	1.42669716	1.8536426
$f(x)_{30}$	2.814168743	2.629049484	2.93093628	2.5140477	1.475258099

in 19 of the 26 selected functions, indicating a 73% of the employed test set and the most consistent *SR* percentage in 23 of the 26 selected test functions. The proposed GIBBS methodology represents a notorious improvement and presents a solid robustness and consistency in the results obtained for all the executions employed for the evaluation. Complementary to this, the results from Table 3.6 exhibit that the GIBBS methodology outperforms its competitors, representing a more viable alternative for initializing the candidate solutions to get a better performance on the optimization process.

3.7.4 Evaluation of Convergence Graphs

Complementary to the quantitative analysis of the results of solution quality and convergence speed of the proposed GIBBS method, this section presents a graphical analysis of the convergence of the population to the optimal solution, starting from the initial configuration of the algorithm until the end of the evolutionary process, illustrating the process performed by the metaheuristic algorithm to reach the optimal solution to different optimization problems. Since each of the methods tested was run a total of 30 times for each function evaluated, the plots have been formed from the median of the results obtained. This approach provides a more robust and representative view of the performance of each algorithm, minimizing the impact of outliers or variations caused by randomness.

In Fig. 3.8, convergence plots for four objective-representative functions: (a) $f(x)_2$, (b) $f(x)_9$, (c) $f(x)_{10}$, and (d) $f(x)_{11}$. The graphs of the convergence process are presented for a total of 6 representative functions, which have meant a considerable improvement in comparison with the methods found in the literature. These graphs provide a notorious panorama for the comparison of convergence ranges between the objective functions, indicating the efficiency of each method in reaching an optimal solution (Table 3.7).

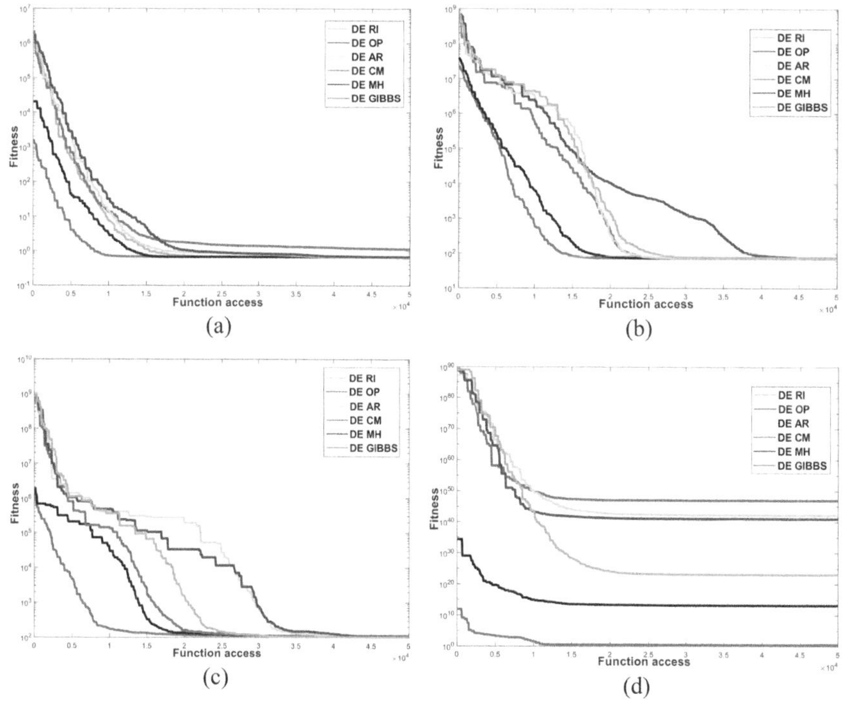

Fig. 3.8 Convergence graphs for a representative set of 4 representative functions: **a** $f(x)_2$, **b** $f(x)_9$, **c** $f(x)_{10}$, and **d** $f(x)_{11}$

Table 3.7 Initial parameters employed for the DE algorithm

Algorithm	Ranking	Mean value
DE_GIBBS	1	5.594594595
DE_MH	2	4.837837838
DE_OP	3	3.59009009
DE_CM	4	3.441441441
DE_RI	5	3.418918919

3.7.5 *Statistical Friedman Ranking Analysis*

The objective of performing a statistical analysis is to compare the uniqueness of a method against others existing in the literature and to highlight the unique properties of a method compared to others. For this analysis, a set of different nonparametric tests were performed to rank the performance of each method based on its characteristics in the different scenarios in which they were tested.

One of the tests performed on the methods is Friedman´s ranking test [31], which is a nonparametric statistical test used to compare and rank different methods across different scenarios and data sets. This test, particularly, is feasible in situations where there is no assumption of normality in the data. Hence, this test provides an alternative to parametric tests such as ANOVA (Analysis of Variance). In Friedman's test, each method is ranked within each data set based on its performance, with the best-performing method receiving first place, second with second place, and so on. In cases where two methods exhibit similar performance, they are assigned the average of the ranks they occupy.

The statistical test of the performance of the methods compared in this study is the Friedman rank test. This test validates the results obtained by the methods compared using the average (AB) presented in Table 3.1.

Table 3.4 shows the results of the Friedman test comparing the effectiveness of each method. The DE_GIBBS method performed superiorly in the experiments.

3.8 Conclusions

Population initialization highlights an important stage for the evolutionary process of all metaheuristic algorithms, not only by giving the set of possible initial solutions with which the optimization work will be performed, but also because it can give from an early stage of the algorithm an overview about the regions with possible optimal solutions, which deserve to be exploited to obtain a global optimal solution, as well as to avoid areas with poor performance for the optimization problem. The relevance becomes more noticeable when increasing the number of dimensions for different optimization problems with various characteristics that increase the difficulty of the problem at hand.

Different initial distributions of individuals proposed in the literature decrease the algorithm's performance, presenting problems with a higher level of complexity as the dimensions increase. These problems include the clustering of individuals in specific zones and seriously affect the exploration of regions with a higher potential to find the global optimum.

In this chapter, a new method for initializing populations based on the Gibbs sampling technique was presented, which allows the generation of individuals with a low degree of correlation, which favors the dispersion of individuals even in problems of high dimensionality, from samples approximating a joint probability distribution. The low correlation feature is particularly beneficial since it prevents the concentration of initial solutions in certain regions of the search space, which is a common problem in high-dimensionality problems. Attacking this problem results in an improvement in the exploration capabilities of metaheuristic algorithms, especially in complex multidimensional spaces.

The effectiveness of this proposed technique has been tested by applying it to the Differential Evolution (DE) algorithm, tested on a set of 30 benchmark functions, and the results obtained have been compared against other initialization methods in the literature applied to the same algorithm with the same parameters to ensure a fair comparison.

The results obtained and presented in the experimentation have shown that the proposed algorithm is able to generate a superior set of initial solutions, resulting in an effective solution to the proposed problem, applying a correct probabilistic sampling, corroborating how a correct distribution of solutions improves the performance of a metaheuristic algorithm. The robust foundation of the initial solutions plays a critical role for the overall performance of the algorithm, particularly in consistently identifying the global optimal solution. Unlike other methods that face problems with increasing complexity, this algorithm maintains its effectiveness by navigating the expanded search space.

Appendix: Employed Test Functions

See Table 3.8.

Table 3.8 Mathematical representation of employed test functions

	Name	Global minimum	Boundaries	Mathematical function
$f(x)_1$	Ackley	$f(\mathbf{x}^*) = 0$ $\mathbf{x}^* = (0, \ldots, 0)$	$[-30, 30]^d$	$f(x) = -20\exp\left(-0.2\sqrt{\frac{1}{d}\sum_{i=1}^{d} x_i^2}\right) - exp\left(\frac{1}{d}\sum_{i=1}^{d}\cos(2\pi x_i)\right) + 20 + \exp(1)$
$f(x)_2$	Dixon-Price	$f(\mathbf{x}^*) = 0$ $\mathbf{x}^* = 2^{-\frac{2^i-2}{2^i}}$ for $i = 1, \ldots, n$	$[-10, 10]^d$	$f(x) = (x_1 - 1)^2 + \sum_{i=1}^{d} i\left(2x_i^2 - x_{i-1}\right)^2$
$f(x)_3$	Griewank	$f(\mathbf{x}^*) = 0$ $\mathbf{x}^* = (0, \ldots, 0)$	$[-600, 600]^d$	$f(x) = \sum_{i=1}^{d} \frac{x_i^2}{4000} - \prod_{i=1}^{d} \cos\left(\frac{x_i}{\sqrt{i}}\right) + 1$
$f(x)_4$	Infinity	$f(\mathbf{x}^*) = 0$ $\mathbf{x}^* = (0, \ldots, 0)$	$[-1, 1]^d$	$f(x) = \sum_{i=1}^{d} x_i^6 sen(x_i + 2)$
$f(x)_5$	Levy	$f(\mathbf{x}^*) = 0$ $\mathbf{x}^* = (1, \ldots, 1)$	$[-10,10]^d$	$f(x) = sin^2(\pi\omega_1) + \sum_{i=1}^{d-1}(\omega_1 - 1)^2\left[1 + 10sin^2(\pi\omega_i + 1)\right] + (\omega_d - 1)^2\left[1 + sin^2(2\pi\omega_d)\right]$ $\omega_i = 1 + \frac{x_i - 1}{4}$
$f(x)_6$	Mishra 1	$f(\mathbf{x}^*) = 2$ $\mathbf{x}^* = (0, \ldots, 0)$	$[0,1]^d$	$f(x) = \left(1 + \left(d - \sum_{i=1}^{d-1} x_i\right)\right)^{d - \sum_{i=1}^{n-1} x_i}$
$f(x)_7$	Mishra 2	$f(\mathbf{x}^*) = 2$ $\mathbf{x}^* = (1, \ldots, 1)$	$[0,1]^d$	$f(x) = \left(1 + \left(d - \sum_{i=1}^{d-1} \frac{x_i + x_{i+1}}{2}\right)\right)^{d - \sum_{i=1}^{n-1} \frac{x_i + x_{i+1}}{2}}$
$f(x)_8$	Multimodal	$f(\mathbf{x}^*) = 0$ $\mathbf{x}^* =$ $(-1, \ldots, -1)$	$[-10,10]^d$	$f(x) = \sum_{i=1}^{d} \lvert x_i \rvert \prod_{i=1}^{d} \lvert x_i \rvert$

(continued)

Table 3.8 (continued)

	Name	Global minimum	Boundaries	Mathematical function
$f(x)_9$	Penalty 1	$f(\mathbf{x}^*) = 0$ $\mathbf{x}^* = (-1, \ldots, -1)$	$[-50{,}50]^d$	$f(x) = \frac{\pi}{d} \times \left\{10\sin^2(\pi\varphi_1) + \sum_{i=1}^{d-1}(\varphi_i - 1)^2\left[1 + 10\sin^2(\pi\varphi_{i+1})\right] + (\varphi_d - 1)^2\right\} + \sum_{i=1}^{d} u(x_i, a, k, m)$ $\varphi_i = 1 + \frac{1}{4}(x_i + 1),\ u(x_i, a, k, m) = \begin{cases} k(x_i - a)^m \ if \ x_i > a \\ 0 if \ -a \leq x_i \\ k(-x_i - a)^m \ if \ x_i < a \end{cases} ;\ a = 10, k = 100, m = 4$
$f(x)_{10}$	Penalty 2	$f(\mathbf{x}^*) = 0$ $\mathbf{x}^* = (1, \ldots, 1)$	$[-50{,}50]^d$	$f(x) = 0.1 \times \left\{\sin^2(3\pi x_1) + \sum_{i=1}^{d-1}(x_i - 1)^2\left[1 + \sin^2(3\pi x_{i+1})\right] + (x_d - 1)^2[1 + \sin^2(2\pi x_d)]\right\} + \sum_{i=1}^{n} u(x_i, a, k, m)$ $u(x_i, a, k, m) = \begin{cases} k(x_i - a)^m \ if \ x_i > a \\ 0 \ if \ -a \leq x_i \\ k(-x_i - a)^m \ if \ x_i < a \end{cases} ;\ a = 5, k = 100, m = 4$
$f(x)_{11}$	Perm 2	$f(\mathbf{x}^*) = 0$ $\mathbf{x}^* = (1, 1/2, \ldots, 1/n)$	$[-d, d]^d$	$f(x) = \sum_{i=1}^{d}\left[\sum_{j=1}^{d}(j^i + 10)\left(x_j^i - \frac{1}{j^i}\right)\right]^2$
$f(x)_{12}$	Plateau	$f(\mathbf{x}^*) = 30$ $\mathbf{x}^* = (0, \ldots, 0)$	$[-5.12, 5.12]^d$	$f(x) = 30 + \sum_{i=1}^{d} \lvert x_i \rvert$
$f(x)_{13}$	Powell	$f(\mathbf{x}^*) = 0$ $\mathbf{x}^* = (0, \ldots, 0)$	$[-4, 5]^d$	$f(x) = \sum_{i=1}^{\frac{d}{4}}[(x_{4i-3} + 10x_{4i-2})^2 + 5(x_{i-1} + x_{4i})^2 + (x_{4i-2} + 2x_{4i-1})^4 + 10(x_{4i-3} + x_{4i})^4]$
$f(x)_{14}$	Qing	$f(\mathbf{x}^*) = 0$ $\mathbf{x}^* = (0, \ldots, 0)$	$[-1.28, 1.28]^d$	$f(x) = \sum_{i=1}^{d}(x_i^2 - i)^2$

(continued)

Table 3.8 (continued)

	Name	Global minimum	Boundaries	Mathematical function
$f(x)_{15}$	Quartic	$f(\mathbf{x}^*) = 0$ $\mathbf{x}^* = (-1, \ldots, -1)$	$[-10, 10]^d$	$f(x) = \sum_{i=1}^{d} ix_i^4 + rand[0,1]$
$f(x)_{16}$	Quintic	$f(\mathbf{x}^*) = 0$ $\mathbf{x}^* = (0, \ldots, 0)$	$[-5.12, 5.12]^d$	$f(x) = \sum_{i=1}^{d} \lvert x_i^5 - 3x_i^4 + 4x_i^3 + 2x_i^2 - 10x_i - 4 \rvert$
$f(x)_{17}$	Rastringin	$f(\mathbf{x}^*) = 0$ $\mathbf{x}^* = (1, \ldots, 1)$	$[-5,10]^d$	$f(x) = 10d + \sum_{i=1}^{d} [x_i^2 - 10cos(2\pi x_i)]$
$f(x)_{18}$	Rosenbrock	$f(\mathbf{x}^*) = 0$ $\mathbf{x}^* = (0.5, \ldots, 0.5)$	$[-100,100]^d$	$f(x) = \sum_{i=1}^{d} 100(x_{i+1} - x_i^2)^2 + (x_i - 1)^2$
$f(x)_{19}$	Schweffel 21	$f(\mathbf{x}^*) = 0$ $\mathbf{x}^* = (0, \ldots, 0)$	$[-100,100]^d$	$f(x) = max\{\lvert x_i \rvert, 1 \le i \le d\}$
$f(x)_{20}$	Schweffel 22	$f(\mathbf{x}^*) = 0$ $\mathbf{x}^* = (0, \ldots, 0)$	$[-100,100]^d$	$f(x) = \sum_{i=!}^{d} \lvert x_i \rvert + \prod_{i=1}^{d} \lvert x_i \rvert$
$f(x)_{21}$	Step	$f(\mathbf{x}^*) = 0$ $\mathbf{x}^* = (0, .., 0)$	$[-100,100]^d$	$f(x) = \sum_{i=1}^{d} \lvert x_i^2 \rvert$
$f(x)_{22}$	Styblinski-Tang	$f(\mathbf{x}^*) = -39.1659n$ $\mathbf{x}^* = (-2.9, .., -2.9)$	$[-5,5]^d$	$f(x) = \frac{1}{2} \sum_{i=1}^{d} (x_i^4 - 16x_i^2 + 5x_i)$
$f(x)_{23}$	Zakharov	$f(\mathbf{x}^*) = 0$ $\mathbf{x}^* = (0, \ldots, 0)$	$[-5,10]^d$	$f(x) = \sum_{i=1}^{d} x_i^2 + \left(\sum_{i=1}^{d} 0.5ix_i\right)^2 + \left(\sum_{i=1}^{d} 0.5ix_i\right)^4$
$f(x)_{24}$	Rothyp	$f(\mathbf{x}^*) = 0$ $\mathbf{x}^* = (0, \ldots, 0)$	$[-65.536, 65.536]^d$	$f(x) = \sum_{i=1}^{d} \sum_{j=1}^{i} x_j^2$

(continued)

Table 3.8 (continued)

	Name	Global minimum	Boundaries	Mathematical function
$f(x)_{25}$	Schweffel 2	$f(\mathbf{x}^*) = 0$ $\mathbf{x}^* = (0, \ldots, 0)$	$[-100,100]^d$	$f(x) = \sum_{i=1}^{d} \left(\sum_{j=1}^{i} x_i\right)^2$
$f(x)_{26}$	Sphere	$f(\mathbf{x}^*) = 0$ $\mathbf{x}^* = (0, \ldots, 0)$	$[-100,100]^d$	$f(x) = \sum_{i=1}^{d} x_i^2$
$f(x)_{27}$	Sum squares	$f(\mathbf{x}^*) = 0$ $\mathbf{x}^* = (0, \ldots, 0)$	$[-10,10]^d$	$f(x) = \sum_{i=1}^{d} ix_i^2$
$f(x)_{28}$	Rastringin + Schweffel 22 + Sphere	$f(\mathbf{x}^*) = 0$ $\mathbf{x}^* = (0, \ldots, 0)$	$[-100,100]^d$	$f(x) = \left[10d + \sum_{i=1}^{d}\left[x_i^2 - 10cos(2\pi x_i)\right]\right] + \left[\sum_{i=!}^{d}\lvert x_i\rvert + \prod_{i=1}^{d}\lvert x_i\rvert\right] + \left[\sum_{i=1}^{d} x_i^2\right]$
$f(x)_{29}$	Ackley + Penalty 2 + Rosenbrock + Schweffel 22	$f(\mathbf{x}^*) = (1.1n) - 1$ $\mathbf{x}^* = (0, \ldots, 0)$	$[-100,100]^d$	$f(x) = f(x)_1 + f(x)_{10} + f(z)_{18} + f(x)_{20}$
$f(x)_{30}$	Ackley + Griewank + Rastringin + Rosenbrock + Schweffel 22	$f(\mathbf{x}^*) = n - 1$ $\mathbf{x}^* = (0, \ldots, 0)$	$[-100,100]^d$	$f(x) = f(x)_1 + f(x)_3 + f(x)_{17} + f(x)_{18} + f(x)_{20}$

References

1. B. Guenin, J. Koenemann, *A Gentle introduction to optimization*, vol. 1, 1st edn. (Cambridge University Press, Waterloo, Ontario, 2014)
2. X.-S. Yang, *Optimization Techniques and Applications with Examples*, 1st edn., vol. 1. (Wiley, United States of America, 2018)
3. S. Carbas, A. Toktas, and D. Ustun, *Nature-Inspired Metaheuristic Algorithms for Engineering Optimization Applications*. Singapore: Springer Singapore, 2021. https://doi.org/10.1007/978-981-33-6773-9.
4. E. Cuevas, A. Rodríguez, *Metaheuristic Computation with MATLAB®*, First edn. (Chapman and Hall/CRC Press, Boca Raton, 2020). https://doi.org/10.1201/9781003006312
5. J.O. Agushaka, A.E. Ezugwu, L. Abualigah, S.K. Alharbi, H.A.E.-W. Khalifa, Efficient initialization methods for population-based metaheuristic algorithms: a comparative study. Arch. Comput. Methods Eng. **30**(3), 1727–1787 (2023). https://doi.org/10.1007/s11831-022-09850-4
6. M. Sarhani, S. Voß, R. Jovanovic, Initialization of metaheuristics: comprehensive review, critical analysis, and research directions. Int. Trans. Oper. Res. **30**(6), 3361–3397 (2023). https://doi.org/10.1111/itor.13237
7. P.A. Digehsara, S.N. Chegini, A. Bagheri, M.P. Roknsaraei, An improved particle swarm optimization based on the reinforcement of the population initialization phase by scrambled Halton sequence. Cogent Eng. **7**(1), 1737383 (2020). https://doi.org/10.1080/23311916.2020.1737383
8. S.J. Mousavirad, A.A. Bidgoli, S. Rahnamayan, Tackling deceptive optimization problems using opposition-based DE with center-based Latin hypercube initialization, in *2019 14th International Conference on Computer Science & Education (ICCSE)*, IEEE, August 2019 (2019), pp. 394–400. https://doi.org/10.1109/ICCSE.2019.8845360
9. I.H. Sloan, Lattice methods for multiple integration. J. Comput. Appl. Math. **12–13**, 131–143 (1985). https://doi.org/10.1016/0377-0427(85)90012-3
10. Q. Li, Y. Bai, W. Gao, Improved initialization method for metaheuristic algorithms: a novel search space view. IEEE Access **9**, 121366–121384 (2021). https://doi.org/10.1109/ACCESS.2021.3073480
11. S. Rahnamayan, H.R. Tizhoosh, M.M.A. Salama, A novel population initialization method for accelerating evolutionary algorithms. Comput. Math. Appl. **53**(10), 1605–1614 (2007). https://doi.org/10.1016/j.camwa.2006.07.013
12. M.F. Ahmad, N.A.M. Isa, W.H. Lim, K.M. Ang, Differential evolution with modified initialization scheme using chaotic oppositional based learning strategy. Alex. Eng. J. **61**(12), 11835–11858 (2022). https://doi.org/10.1016/j.aej.2022.05.028
13. W. Pan, K. Li, M. Wang, J. Wang, B. Jiang, Adaptive randomness: a new population initialization method. Math. Probl. Eng. **2014**(1) (2014). https://doi.org/10.1155/2014/975916
14. E. Cuevas, H. Escobar, R. Sarkar, H.F. Eid, A new population initialization approach based on Metropolis-Hastings (MH) method. Appl. Intell. **53**(13), 16575–16593 (2023). https://doi.org/10.1007/s10489-022-04359-6
15. Q. Li, S.-Y. Liu, X.-S. Yang, Influence of initialization on the performance of metaheuristic optimizers. Appl. Soft Comput. **91**, 106193 (2020). https://doi.org/10.1016/j.asoc.2020.106193
16. B. Kazimipour, X. Li, A.K. Qin, A review of population initialization techniques for evolutionary algorithms," in *2014 IEEE Congress on Evolutionary Computation (CEC)*, IEEE, July 2014 (2014), pp. 2585–2592. https://doi.org/10.1109/CEC.2014.6900618
17. G. Casella, E.I. George, Explaining the Gibbs sampler. Am. Stat. **46**(3), 167–174 (1992). https://doi.org/10.1080/00031305.1992.10475878
18. L. Kocis, W.J. Whiten, Computational investigations of low-discrepancy sequences. ACM Trans. Math. Softw. **23**(2), 266–294 (1997). https://doi.org/10.1145/264029.264064
19. G.E. Fasshauer, *Meshfree Approximation Methods with Matlab*, vol. 6 (World Scientific, 2007). https://doi.org/10.1142/6437

20. M.D. Mckay, R.J. Beckman, W.J. Conover, A comparison of three methods for selecting values of input variables in the analysis of output from a computer code. Technometrics **42**(1), 55–61 (2000). https://doi.org/10.1080/00401706.2000.10485979
21. A.M. Elsawah, K.-T. Fang, Y.H. Deng, Some interesting behaviors of good lattice point sets. Commun. Stat. Simul. Comput. **50**(11), 3650–3668 (2021). https://doi.org/10.1080/03610918.2019.1628988
22. K.-T. Fang, X. Ke, A.M. Elsawah, Construction of uniform designs via an adjusted threshold accepting algorithm. J. Complex. **43**, 28–37 (2017). https://doi.org/10.1016/j.jco.2017.05.002
23. C.-H. Chou, J.-N. Chen, Genetic algorithms: initialization schemes and genes extraction, in *Ninth IEEE International Conference on Fuzzy Systems. FUZZ- IEEE 2000 (Cat. No.00CH37063)*, IEEE, pp. 965–968. https://doi.org/10.1109/FUZZY.2000.839167
24. A. Bedri Ozer, CIDE: chaotically initialized differential evolution. Expert Syst Appl. **37**(6), 4632–4641 (2010). https://doi.org/10.1016/j.eswa.2009.12.045
25. H.R. Tizhoosh, Opposition-based learning: a new scheme for machine intelligence, in *International Conference on Computational Intelligence for Modelling, Control and Automation and International Conference on Intelligent Agents, Web Technologies and Internet Commerce (CIMCA-IAWTIC'06)*, IEEE, pp. 695–701. https://doi.org/10.1109/CIMCA.2005.1631345
26. P.-E. Danielsson, Euclidean distance mapping. Comput. Graph. Image Process. **14**(3), 227–248 (1980). https://doi.org/10.1016/0146-664X(80)90054-4
27. H.-O. Peitgen, H. Jürgens, D. Saupe, *Chaos and Fractals* (Springer, New York, 2004). https://doi.org/10.1007/b97624
28. S. Geman, D. Geman, Stochastic relaxation, Gibbs distributions, and the Bayesian restoration of images. IEEE Trans. Pattern Anal. Mach. Intell. **PAMI-6**(6), 721–741 (1984). https://doi.org/10.1109/TPAMI.1984.4767596
29. R. Storn, K. Price, Differential evolution – a simple and efficient heuristic for global optimization over continuous spaces. J. Glob. Optim. **11**(4), 341–359 (1997). https://doi.org/10.1023/A:1008202821328
30. K. Hussain, M.N. Mohd Salleh, S. Cheng, R. Naseem, Common benchmark functions for metaheuristic evaluation: a review. JOIV: Int. J. Inf. Vis. **1**(4–2), 218–223 (2017). https://doi.org/10.30630/joiv.1.4-2.65
31. M. Friedman, A comparison of alternative tests of significance for the problem of m rankings. Ann. Math. Stat. **11**, 86–92 (1940)

Chapter 4
Metaheuristic Optimization with Dynamic Strategy Adaptation: An Evolutionary Game Theory Approach

4.1 Game Theory

Game theory is a branch of applied mathematics that uses models to study interactions in formalized stimulus structures called "games." Game theory has become an essential tool in economics and business administration, contributing to a more accurate understanding of human decision-making behavior. The research area of this branch focuses on studying optimal strategies and the expected and observed behavior of people in games [1].

Initially developed as a tool for understanding behavior in economics, game theory has spread to many other fields, such as biology, computer science, sociology, political science, psychology, and philosophy. This area underwent significant development and was first formalized in the work of John von Neumann and Oskar Morgenstern during the Cold War, mainly through its application to military strategy. Since the 1970s, game theory has been applied to animal behavior, including the evolution of species through natural selection.

In the most basic sense of this theory, the game is a conflict situation in which the conflict benefits of people or organizations dominate. In this context, when making a decision, it affects the determination of the determination that others will give. Therefore, the outcomes of the conflict are determined by all decisions approved by all units [2]. Game theory shows that it is necessary to have a reasonable way of playing any game, especially in the case of many cons and second intentions. Therefore, for example, mutual expectations for opposite intentions that appear in games such as chess or poker, leading to infinite theoretical chains, can also be transferred to resolutions of real and complicated conflicts. In this way, and as discussed, people interacting in conflicts will receive outcomes that, in a sense, depend entirely on that interaction [3].

E. Cuevas et al., *Advanced Metaheuristics: Novel Approaches for Complex Problem Solving*, Studies in Computational Intelligence 1210,
https://doi.org/10.1007/978-3-031-89284-4_4

4.2 Classical Game Theory

When applying these models, it is essential to remember that game theory is an abstract representation of reality and that the actions of an individual will somehow affect the state of other players, so it is essential to consider the assumptions when evaluating it because it is almost impossible to represent what happens in real life using a model accurately. In this regard, game theory is based on the following assumptions [4]:

- Each player has two or more clearly defined options to choose from, called "moves."
- Any combination of moves available to the players leads them to a well-defined end state (win, lose, or surrender), thus ending the game.
- Each end state grants special rewards to each player.
- Each player has a complete understanding of the game and the opponent, which means that the player understands the rules of the game in detail, as well as their preferences and opinions, what other players gain, and what each player can or cannot do to their advantage.
- All players are rational; this means that each player has two options and will choose the one that represents the most significant interest or utility.

Thus, game theory is a general theory of rational behavior that applies when two or more players have freedom and a limited number of courses of action, leading to a clear outcome. Victory and defeat are represented by a numerical payoff associated with each combination of courses of action and for each player. The players are well aware of the rules of the game and are rational because each player optimizes their benefit.

4.2.1 Types of Games

In the analysis of Gibbons' game theory in 1996, the existence of four types of games is pointed out, which are static games, dynamic games, complete games and incomplete games [5]. Static games with complete information are defined on the basis of the existence of two players, each of whom simultaneously chooses and executes a certain option from a menu of possible alternatives. In the end, each person gets a reward.

Gibbons (1996), when speaking of static games with incomplete information, also called Bayesian games, defines them in the same way as the previous game but takes into account that at least one of the players plays uncertainty about the perceived benefit or payoff function of the other player.

In the case of a dynamic game with complete information, the author assumes that the first player chooses an option from a set of possible alternatives, and the second player observes the decision chosen by the first player and then performs his action from the set of possible alternatives. After performing each action, each player involved receives a reward.

In contrast, a dynamic game with incomplete information means that one player has private information and the opponent does not, leading to the latter following the player's actions with the information. This type of game is called a "signaling game" because signals sent by players who know confidential information will trigger actions by their opponents based on the information received.

However, when reviewing the current literature on game classification, it can be observed that several authors, including Binmore [6] and Shubik [7], classify games as cooperative and non-cooperative. There are two main types of games, or, in other words, two different approaches to game analysis: cooperative and non-cooperative. The collaborative approach explores the possibility that some or all of the participating parties reach an agreement on the decisions that each party would make in the absence of prior agreement.

It is possible that some players will try to reach binding agreements, which we call coalitions. If the only possible action of a player in a game is to cooperate with others, then the game will focus on examining the results that each player's alliance can achieve. We talk about analyzing how a group of players can act, taking care of the collective behavior and not limiting ourselves to the individual actions of each member of the alliance. In such a situation, it is necessary to take into account the factors that will influence what alliance will be formed and how the resulting benefits will be distributed.

Non-cooperative games are games in which players make decisions independently based on their interests, which does not exclude that, in some cases, decision-making may benefit everyone, as is the case in cooperative games. Non-cooperative games are often analyzed in a way that seeks to predict players' strategies and payoffs, as well as to find Nash equilibria. This contrasts with cooperative game theory, which focuses on predicting which groups players will form coalitions with, what actions the groups will take, and the collective benefits they may bring.

Cooperative game theory does not analyze the strategic negotiations that take place within each coalition and influence the distribution of collective benefits among the members of that coalition. Non-cooperative game theory provides a low-level approach because it models all the procedural details of the game. In contrast, cooperative game theory only describes the structure, strategy, and outcomes of the alliance. In this sense, non-cooperative game theory is more comprehensive than cooperative game theory [8].

When there is an arbitrator to enforce the contract, the contract does not fall under the non-cooperative game theory. However, sufficient assumptions can be made to account for all possible strategies that the players may adopt in relation to the arbitration. Another way is to describe the arbitrator as part of the contract and model the related processes and payments accordingly.

Therefore, it is desirable that all games take place in a non-cooperative framework. However, in many cases, there is insufficient information to model the formal procedures available to those involved in strategic negotiations accurately. It may also happen that the resulting model is too complex to provide a practical tool in the real world. In such cases, cooperative game theory offers a simplified approach that allows one to analyze the game as a whole without having to make assumptions about bargaining power [9].

4.2.2 The Prisoner's Dilemma

In classical game theory, the most common example of application is the prisoner's dilemma, where it is important to know the concept of the Nash equilibrium.

Nash equilibrium, Cournot equilibrium, Cournot-Nash equilibrium or fear equilibrium is a game theory solution concept for games with two or more players that assumes that [10]:

- Each player knows and uses their best strategy.
- Each knows the strategies of the others.

As a result, each player gains nothing by changing their strategy while the others keep the same. In this way, each player makes the best "move," taking into account the moves of the other players. In other words, a Nash equilibrium is a situation in which all players have adopted and know that they have adopted a strategy that maximizes their profits, given the strategies of the others. Therefore, no player has any incentive to change his strategy.

It is important to remember that Nash equilibrium is not about achieving the best collective outcome for the participants but only the best outcome for each of them, considered separately. It is quite possible that the outcome would be better for everyone if the players coordinated their actions in some way [11].

A problem with Nash equilibrium is that it does not necessarily lead to a Pareto efficient situation. The initial analysis of this game was based on a scenario in which two people who commit an armed bank robbery are interrogated in separate rooms; however, the stolen money was not in their hands, so the police could only charge them with illegal possession of weapons without any other evidence. Therefore, during a separate interrogation, each of them will have the opportunity to admit their guilt, indicate the involvement of another prisoner or deny their participation in the attack. However, the police can offer them a solution and, using an appropriate incentive system, force them to admit their involvement in the crime, reveal the truth and convict them. As seen below, a proper police recommendation can lead to individual rationality and selfishness in decision-making that can often go against the collective interests of these entities [12]. To demonstrate this, consider the Prisoner's Dilemma game. This game makes us realize that maintaining cooperation is extremely difficult. Individuals often do not cooperate, and their decisions do not necessarily lead to the common good (Fig. 4.1).

		Prisoner B	
		Confess	Keep silent
Prisoner A	Confess	Both are sentenced to 6 years.	Prisoner A is released, Prisoner B is sentenced to 10 years.
	Keep silent	Prisoner B is released, Prisoner A is sentenced to 10	Both are sentenced to 1 year.

Fig. 4.1 Prisoner's dilemma reward matrix

4.3 Evolutionary Game Theory

Evolutionary game theory (EGT) is the application of population genetics-based models of gene frequency change in populations to game theory. It differs from classical game theory in that it focuses on the dynamics of strategy rather than its equilibria. Evolutionary game theory began with the problem of explaining the ritual behavior of animals in conflict situations [13].

Maynard Smith realized that the evolutionary version of game theory did not require players to act rationally but only strategically. The outcome of the game shows how good this strategy is, as well as how evolution tests alternative strategies in terms of their ability to survive and reproduce [14].

The success of a strategy depends on how well it performs in the presence of competing strategies (including its own) and how often these strategies are used. Participants try to create as many copies of themselves as possible, and rewards are expressed in fitness units. It is always a multiplayer game with many opponents. The rules cover the dynamics of the copying mechanism, in other words, how the fittest players will create more copies of themselves in the population and how the fittest players will be removed from the replication mechanism equation [15]. Replicator dynamics models inheritance rather than mutation and assumes asexual reproduction for simplicity. The game is restarted without end conditions. Outcomes include population dynamics, strategy success, and any equilibrium achieved. Unlike classical game theory, players do not choose their strategy and cannot change it: they are born with a strategy, and their offspring inherit it.

Evolutionary game theory analyzes Darwinian mechanisms using a systems model that includes three main components: population dynamics, games and replication mechanisms. The system process includes four stages [16]:

- The model involves population. The population will show differences between competing individuals. In the model, this capability is represented by the game.
- The game tests each individual's strategy according to the "rules of the game." These rules provide different benefits depending on the fitness unit. Competitive individuals are grouped to compete with other individuals, often in a very heterogeneous population. Combining strategies within a population affects the winning

outcomes by changing the probability of an individual competing in multi-strategy contests. Participants exit the game as a match whose ultimate viability depends on the outcome of the competition, expressed in the reward matrix.

- Depending on the fitness obtained, each member of the population can be copied or eliminated, which is determined by the precise mathematics of the Replicator Dynamics Process. This overall process produces a new generation. Each survivor now has a new physical level, depending on the outcome of the game.
- The new generation replaces the previous one, and the cycle repeats itself. Mixtures of populations can reach an evolutionarily stable state that any mutation strategy cannot attack.

EGT refers to the Darwinian evolutionary process, which includes competition, natural selection, and inheritance. EGT has contributed to the understanding of group selection, sexual selection, altruism, parental care, cooperation, and ecological dynamics. These models have given a sound mathematical basis to many counterintuitive situations in these fields.

Evolutionary games are mathematical objects with different rules, outcomes, and mathematical behavior. Each game represents different challenges that organisms face and strategies they can adopt to survive and reproduce. Different strategies compete according to the rules of a particular game, and mathematics is used to determine the outcomes and behavior.

4.3.1 *The Hawk-Dove Game*

The first game Maynard Smith analyzed was the classic game of chicken, or "Hawk-Dove," as Smith called it. Its purpose is to analyze the Lorenz and Tinbergen problem: the struggle for the common good [3]. The participants can be Hawks or Doves. These are two subgroups or variations of the species with different strategies. The hawk first shows aggression, then starts to fight until it wins or gets hurt (lose). At first, the dove showed aggression, but when confronted with a stronger force, it fled. If no such escalation occurs, the dove will attempt to share resources.

The reward matrix can be seen in the following illustration (Fig. 4.2).

Because resources are assigned a value of V, combat damage has a cost of C:

	Hawk	**Dove**
Hawk	$\frac{v-c}{2}$	v
Dove	0	$\frac{v}{2}$

Fig. 4.2 Reward matrix of the falcon-pigeon game

- If the hawk encounters the dove, it will get all V resources.
- If the hawk meets another hawk, half the time it wins and half the time it loses. Then, its winning average is V/2 minus C/2.
- If the dove meets the Eagle, it will retire and receive nothing, 0.
- If a dove encounters another dove, they share resources and get V/2.

However, the actual reward depends on the probability of encountering a hawk or a dove, which reflects the percentage of hawks and doves in the population at the time of a given competition. This, again, is determined by the results of all previous competitions [17].

Suppose the cost of losing C exceeds the cost of gaining V. In that case, the operation ends up with an evolutionarily stable strategy, a combination of two methods in which the population of hawks is equal to V/C. The population reaches this equilibrium point if any new hawk or dove causes a temporary disruption in the population [18]. The game solution explains why most animal competitions involve only ritual combat and not direct combat. The outcome does not depend on the good behavior of the species, as Lorenz suggests, but only on the participation in the activities of the so-called selfish genes.

4.4 ES-EGT Optimization Algorithm

In this section, we introduce a novel metaheuristic algorithm that dynamically adapts its search strategy during the optimization process to improve its efficiency [19].

4.4.1 Description of the Metaheuristic Algorithm with Dynamic Strategy Adaptation Based on Evolutionary Game Theory (ES-EGT)

The proposed algorithm is built upon Evolutionary Strategies (ES) due to their notable self-adaptive properties. It incorporates mechanisms from Evolutionary Game Theory (EGT) to enhance the algorithm's efficiency. This integration ensures that the adaptation process considers the information from the corresponding agent and knowledge derived from other population members. Furthermore, our approach introduces an innovative modification to the traditional EGT mechanism. Rather than restricting the adaptation process to pairwise evaluations, it also incorporates data from the best-performing individuals in the population, as determined by their scores. This inclusive adaptation method allows for faster convergence to the most efficient dominant strategy, representing a significant advancement in optimization algorithms.

To assess the proposed approach's performance, a comparative analysis was conducted against multiple established metaheuristic algorithms using a comprehensive set of 30 test functions. The results demonstrate that the new methodology outperforms existing techniques, delivering superior solutions and achieving faster convergence rates.

Evolutionary Strategies (ES) [20] are a distinct category of metaheuristic optimization algorithms inspired by natural evolution and biological principles. The central operation in ES is mutation, which serves as the primary mechanism for maintaining diversity and enabling efficient solution space exploration. A defining feature of ES, distinguishing them from other metaheuristic algorithms, is their self-adaptive nature [21].

In ES, algorithm parameters, particularly those controlling mutation rates, evolve alongside candidate solutions. This dynamic adjustment allows the algorithm to adapt its exploration and exploitation balance based on the success or failure of prior generations [22]. Such self-adaptive features make ES particularly effective for continuous optimization problems characterized by rugged and multimodal landscapes of optimal solutions. Despite these strengths, ES designs often treat each agent as independent. Consequently, the individual adaptation mechanism focuses solely on the performance and experiences of each agent within the search space. This approach can result in redundant explorations and misses the potential benefits of coordination among agents to improve the overall strategy's efficiency and effectiveness.

When agents within a population adapt strategies independently, redundant exploration can occur without utilizing information from other members, leading to limited progress. Agents may repeatedly explore the same search regions, applying similar strategies without achieving better solutions. In contrast, collaborative mechanisms that aggregate results from multiple agents can foster a synergistic effect. This approach allows for the development of more coherent and effective strategies by leveraging the diverse experiences and insights of the population. By promoting information sharing and collective efforts, the solution search process becomes significantly more efficient than isolated individual efforts. This collaborative approach ensures a broader exploration of potential solutions and mitigates the risk of stagnation caused by duplicated efforts in the same regions of the search space.

Evolutionary Game Theory (EGT) [16] is a field within mathematical biology and economics that analyzes the dynamics of strategic interactions among individuals or agents within a population. A core component of EGT is the adaptation of strategies within a population [23]. The objective is to expand and promote the most effective strategies while eliminating less effective ones [24]. Within the EGT framework, strategy adaptation considers the performance of individual agents and the influence and effects of other agents within the population.

In this framework, individuals repeatedly interact pairwise, comparing their outcomes [25]. Through these comparisons, the individual with the lower outcome is identified, and its strategy is adapted to more closely align with the plan of the more successful counterpart [26]. Over successive adaptations, the most effective strategy becomes dominant and gradually adopted by most of the population.

While EGT provides a compelling approach to strategy adaptation, its reliance on pairwise interactions often necessitates numerous iterations to identify the optimal strategy within a population [27]. This limitation arises because pairwise comparisons restrict the evaluation scope to two agents at a time, requiring a large number of comparisons to reflect broader population trends adequately. Consequently, multiple interactions and adjustments are needed before the collective performance of the entire population is effectively represented in the dominant strategy. This iterative nature of EGT poses challenges in achieving rapid convergence to optimal solutions.

4.4.2 Implementation of Evolutionary Strategies in the ES-EGT

The ES method involves a series of processes and operations that guide the evolution of individuals (potential solutions) within the algorithm:

1. **Initialization**: The process starts with the creation of an initial population $\{x_1,\ldots, x_N\}$ of N individuals. Each individual $x_i = (x_{i,1},\ldots, x_{i,d})$ represents a potential solution to a particular optimization problem. Each decision variable of the d-th elements is randomly generated within certain bounds. During the initialization phase, each agent x_i is assigned two variables, σ_i and nv_i. σ_i denotes the maximum possible perturbation of the agent's position, and its value is chosen randomly in an interval between *0* and *1*. nv_i, on the other hand, acts as a counter to record how many times the agent has improved its quality within a given time frame. Initially, the value of nv_i is set to *zero*.
2. **Mutation**: An operation in which the original individual x_i changes to position x_i^* by random alteration. Mutation is crucial for maintaining diversity in the population and exploring new parts of the search space. In ES, the degree of mutation σ_i can be adaptive, changing depending on the performance of the algorithm. The mutation operation is modeled as follows:

$$\mathbf{x}_i^* = \mathbf{x}_i + N(0, \sigma_i) \tag{4.1}$$

   where $N(0,\sigma_i)$ represents the normal probability distribution. σ_i represents the standard deviation, which implies the size of the perturbation applied to the original position x_i.
3. **Substitution**: If the quality of x_i^* is higher than that of x_i ($f(x_i^*)<f(x_i)$ in a minimization case), the value of x_i is replaced by x_i^*; otherwise, the value of x_i remains unchanged. If the quality of x_i^* is higher, this implies that the mutation was successful and improved its original value x_i, so the value of nv_i is increased by one ($nv_i = nv_i + 1$).

4. **Strategy adaptation**: In ES, the mutation rate strategy parameter σ_i is adapted as part of the evolutionary process. The adaptation of σ_i is performed, taking into account the 1/5 rule. This rule is based on the principle that the rate of successful mutations should ideally be around 20% (or 1/5) of all mutations. The process of adaptation of the evolutionary strategy according to the 1/5 rule involves the following operations.

   (A) After a number w of generations or iterations, the algorithm evaluates the values of the success rate. The value of w is usually set according to the following formulation:

$$w = \min(d, 30) \tag{4.2}$$

   where d is the number of decision variables of the problem. Considering the value of w and the number nv_i of times that the solution x_i has been improved, the success rate φ_i is determined as follows:

$$\varphi_i = \frac{nv_i}{w} \tag{4.3}$$

   (B) If φ_i exceeds 1/5, it indicates that the mutations are too conservative and could benefit from being more exploratory. In this situation, the mutation rate σ_i is increased. Conversely, if φi is below 1/5, it signifies that the mutations are too aggressive, disrupting the search pattern. Hence, the mutation rate σi is decreased. Despite the fascinating adaptive features of ES, there is a significant limitation in their adaptive mechanism. Strategy modification in these systems relies solely on the individual agent associated with each strategy, neglecting the collaborative input from other agents. This approach to strategy adjustment means that each agent independently attempts to solve the problem using its unique method, functioning in isolation. This isolated approach is clearly less effective, especially when considering the potential advantages of sharing information among agents.

Combining the knowledge of multiple agents could harness the collective experiences encountered during the search process, enriching the solution strategy and potentially leading to more effective problem-solving outcomes. This absence of collaborative integration in evolutionary strategies can be seen as a major limitation in their design.

4.4.3 *Implementation of Evolutionary Game Theory in the ES-EGT*

The adaptation mechanism in EGT is performed by comparing pairs of individuals, taking into account the following steps. In a population consisting of n elements, the adaptation process starts with a randomly chosen individual, x_i, selecting another member, x_j, from the same population ($i,j \in 1,\ldots,n$). They then compare their strategies. If x_i strategy is less efficient than x_j (meaning that $f(x_i) < f(x_j)$), x_i can modify its strategy to be more similar to that of x_j, with the objective of improving its performance. The decision of x_i to modify its strategy is based on a probabilistic mechanism, which is directly related to the difference in strategies between x_i and x_j. This probability-based method promotes exploration in the search for improved strategies. As the process is repeated in multiple iterations, the most successful strategies naturally take hold in the population, while the least effective ones gradually diminish.

EGT provides multiple attractive aspects for strategy adaptation, considering more than just the individual involved. Nonetheless, its approaches often necessitate numerous iterations to identify the optimal strategy within a population. This requirement for multiple iterations arises from EGT's dependence on pairwise comparisons for strategy adaptation. Each iteration involves comparing only two elements, restricting the evaluation of strategies to these pairs. As a result, to ensure that strategy adaptations effectively capture and incorporate the broader performance trends of the majority of agents within the population, a considerable amount of pairwise comparisons is needed. Consequently, before strategies can accurately reflect the collective performance of the entire population, a substantial number of interactions and fine-tuning are essential.

4.4.4 Operation of the ES-EGT

Within the algorithm's framework, each individual's search strategy is determined by two key parameters: its position x_i and the mutation rate σ_i. Consequently, any modification in the strategy directly entails changes in these specific parameters. The significance and functions of these parameters are visually depicted in Fig. 4.1. This figure clearly shows a population of four individuals, each situated at a different position in the search space. It effectively highlights how variations in position and mutation rate uniquely define each individual's search strategy (Fig. 4.3).

The approach detailed in this chapter is divided into two distinct parts. The first part involves the implementation of the standard Evolutionary Strategies (ES) algorithm. During this phase, strategies are adapted based on the individual performance of each agent, focusing on the performance of each agent's strategy in isolation. The second part adopts a more collective focus. Here, strategies are adjusted by incorporating collective information derived from Evolutionary Game Theory (EGT) mechanisms. This segment of the approach utilizes the interactions and relative successes of strategies within the population of agents, enabling more integrated strategy adaptation.

Adaptation using ES: During this phase, we execute the Evolutionary Strategies (ES) algorithm. At each iteration, the algorithm generates a mutant variant x_i^* from each candidate solution x_i according to Eq. 4.1. The performance of this mutant

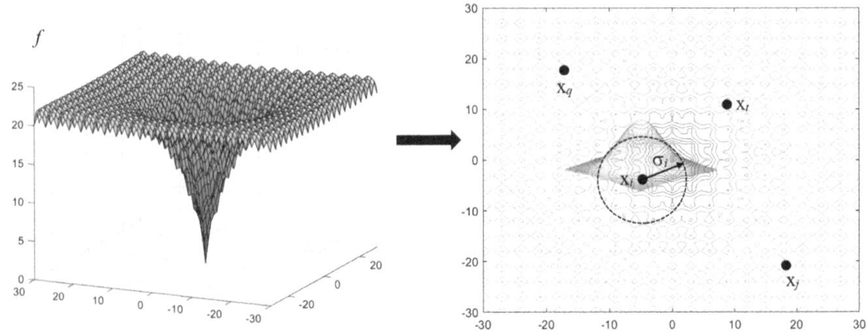

Fig. 4.3 Parameters x_i and σ_i in which the strategy consists for each individual of the population

solution x_i^* is then compared to the original x_i. If x_i^* proves to be better than x_i, two actions are taken: first, a counter nv_i is incremented, tracking the successful mutations; second, the superior mutant solution x_i^* replaces the original x_i. Conversely, if the mutant does not improve the original solution x_i, no changes are made.

The 1/5 Rule: Following a sequence of w iterations, the adaptation of strategies within the algorithm adheres to the 1/5 rule. This essential step utilizes the counter nv_i, which records the frequency of successful mutations for each candidate solution x_i.

The success rate ϕ_i for each candidate solution x_i is determined by analyzing this data. Subsequently, the process incorporates a factor, denoted c, which enables the mutation rate σ_i of each candidate x_i to be adjusted. The 1/5 rule is then systematically applied; if a candidate's mutation success rate exceeds the 1/5 threshold, indicating excessively conservative mutations, σ_i is increased. Conversely, if the success rate ϕ_i is below this threshold, σ_i is decreased, indicating overly aggressive mutations. The following formulation models this adaptive approach guided by the 1/5 rule:

$$\sigma_i = \begin{cases} c\sigma_i & \varphi_i \leq \frac{1}{5} \\ \frac{\sigma_i}{c} & \varphi_i > \frac{1}{5} \end{cases} \tag{4.4}$$

where the factor c is a value between *0* and *1*. In this approach, the value of c has been set at *0.817*.

As is evident, the adaptation of the mutation rate σ_i, and consequently the strategy of each candidate solution x_i, depends exclusively on the performance of the respective agent. This performance is quantitatively reflected by the number of successful mutations nv_i recorded for that agent. In other words, the agent's ability to generate advantageous mutations directly influences how its strategy evolves. If an agent consistently produces advantageous mutations, this indicates the need to adjust its mutation rate, aligning the strategy more closely with its demonstrated success. This approach ensures that the adaptation of the strategy is intimately linked to the performance of each agent, making the process highly tailored and responsive to the specific evolutionary trajectory of each agent.

Adaptation using EGT: During this phase, the modification of strategies involves the incorporation of knowledge from other agents in the population, using the adaptation mechanism of Evolutionary Game Theory (EGT). Traditionally, EGT mechanisms focus on interactions between only two individuals, x_i and x_j. However, in this approach, we have introduced a third fundamental interaction. It involves the agent x_{BEST}, which presents the best performance in terms of the objective function at that time, representing the most successful strategy known at that time. The inclusion of x_{BEST} plays a crucial role. Essentially, it acts as a benchmark or model strategy that others can aspire to or learn from. This lead agent's strategy is considered the best available, given current information and progress within the population. By integrating this element into the strategy adaptation process, we accelerate convergence toward the most effective and dominant strategy. This faster convergence is achieved through a collaborative process in which all actors in the population contribute to and benefit from shared knowledge and successful strategies.

This phase is initiated after the completion of w iterations. In this phase, each individual x_i undertakes a process of comparing its fitness with that of another randomly selected individual x_j. The essence of this comparison lies in evaluating the effectiveness of their strategies, reflected in their fitness levels. If individual x_i has a lower fitness than individual x_j, indicating that his strategy is less effective, a modification is applied. In response to this result, the strategy of individual x_j is adapted to resemble the combination of the strategies used by individual x_j and the current best-performing individual x_{BEST} in the group.

Obtaining distances: In the evaluation process, two different distances are calculated. The first measured distance, $D_{i,j,}$ is between the lowest quality individual x_i and the highest quality individual x_j in the comparison. The second evaluated distance D_{BEST} is between the lowest quality individual x_i and the highest ranked individual x_{BEST} in the population.

$$D_{i,j} = \left|\mathbf{x}_j - \mathbf{x}_i\right| \tag{4.5}$$

$$D_{BEST} = |\mathbf{x}_{BEST} - \mathbf{x}_i| \tag{4.6}$$

$D_{i,\,j}$ provides insight into the difference in performance or efficacy between the strategies employed in the pairwise comparison. D_{BEST} focuses on understanding the spatial or positional difference within the population, providing insight into the distance separating the least effective individual x_i from the most strategically advantageous position x_{BEST}. Together, these two distances provide a comprehensive view of the range and distribution of performance and positioning within the population, which is crucial for strategy evaluation and adaptation.

Influence of the elements x_j and x_{BEST} on the adaptation of the new strategy for *xi*: After the calculation of the distances, we proceed to the calculation of two critical factors, X_n and X_m. These factors are important to determine the degree of influence that the high-quality individual x_j and the best-positioned individual x_{BEST} will exert on the lowest-quality individual x_i.

$$\mathbf{X}_n = \mathbf{x}_j - \sigma_j \cdot D_{i,j} \tag{4.7}$$

$$\mathbf{X}_m = \mathbf{x}_{BEST} - \sigma_{BEST} \cdot D_{BEST} \tag{4.8}$$

Essentially, X_n quantifies the impact of the top-quality individual x_j, while X_m evaluates the influence of the individual x_{BEST} with the most important position in the population. These factors are critical in guiding the adaptation process of the lower-quality individual x_i, dictating the extent to which he must adjust his strategy or position based on the performance and position of the more successful members of the population. By calculating X_n and X_m, a targeted and informed approach to strategy improvement and positional adjustment is facilitated, increasing the overall effectiveness of the optimization process.

Obtaining the new position of x_i^{k+1}: The new position of x_i^{k+1} is computed as a stochastic version of the average position of the two vectors X_n and X_m that consider the influence of the strategies x_j and x_{BEST}. Therefore, the new position of x_i^{k+1} is calculated as follows.

$$\mathbf{x}_i(k+1) = \frac{(\mathbf{X}_n + \mathbf{X}_m)}{2} \sin(2 \cdot \pi \cdot R) e^a \tag{4.9}$$

where R represents a random value uniformly distributed between 0 and 1. a corresponds to a factor whose value decreases as the number of iterations progresses according to the following model:

$$a = 1 - \frac{k \cdot w}{maxiter} \tag{4.10}$$

where k represents the current iteration and $maxiter$ is the maximum number of iterations the algorithm executes.

Equation 4.9 introduces the term $sin(2 \cdot \pi \cdot R)$, which plays an important role in the equation by generating a random value that falls between 1 and -1, with the specific intention of perturbing the mean value. However, the behavior of the term differs markedly from what would be expected from a uniformly distributed probability function. This discrepancy arises due to the incorporation of the $sin(\bullet)$ function in determining its value. As a result of utilizing the sin function, the likelihood of obtaining the same positive or negative perturbation value is effectively doubled compared to a uniform distribution. This characteristic of the term creates a distinctly different effect, as it modifies the distribution and frequency of perturbations applied to the mean value, impacting the overall dynamics and results of the equation in a manner that deviates from standard uniform probability distributions.

For a metaheuristic algorithm to effectively navigate a search strategy, it must adeptly balance exploration and exploitation within the search space. Exploration involves discovering new and promising regions within the search space, while exploitation focuses on refining and optimizing the solutions already identified. The literature in this field [28–30] often suggests beginning the search process with

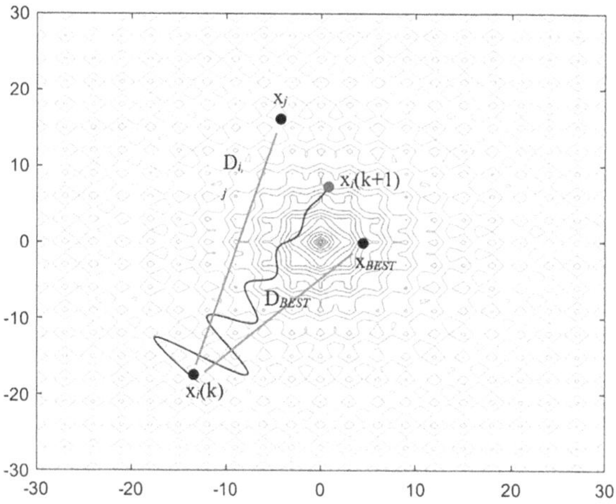

Fig. 4.4 Trajectory resulting from the EGT on an agent x_i

exploration, followed by a shift to exploitation. It is typically recommended that the distribution between these two processes be around 10–20% for exploration and 80–90% for exploitation. This balance is achieved in Eq. 4.7 by the exponential function e^a. This function is designed to vary its output based on the number of iterations. During the initial 20% of the search process, it generates relatively large values, resulting in significant perturbations. These substantial perturbations are essential for exploration, as they enable the algorithm to explore new areas within the search space. Conversely, as the search progresses, the exponential function smoothly decays, ensuring lower values for the remaining 80% of the iterations. These lower values lead to only slight perturbations, which are integral to the exploitation phase. During exploitation, the focus is on fine-tuning and gradually improving the solutions already discovered. This gradual decrease in function output facilitates the transition from broad exploration to focused exploitation, aligning with the ideal operational dynamics for an effective search strategy (Fig. 4.4).

Strategy adaptation: The decision for x_i to change its strategy is based on a probabilistic mechanism, which is directly related to the quality of the original x_i. Therefore, if a candidate solution does not meet the established probabilistic rule, its newly generated value x_i^{k+1} is not accepted and, consequently, is discarded. As a result, the candidate solution x_i remains as it was originally, without incorporating the new value. The process of applying this probabilistic rule begins with a systematic ranking of all solutions in the population $\{x_1,\ldots,x_N\}$ according to their target values. In this ranking system, the solutions are arranged such that the best-performing solution is assigned the highest rank, denoted as number one. Conversely, the solution with the least favorable performance is assigned the rank N, which represents the lowest position in this hierarchy. This ranking method is fundamental, as it provides a clear and structured way to evaluate the relative performance of each solution, forming the basis for the probabilistic decision on whether to retain or discard new values during the optimization process.

The second step consists of assigning the probability of a strategy change Pr_i for agent x_i, which is defined by the following expression:

$$Pr_i = \frac{Rank(\mathbf{x}_i)}{N} \tag{4.11}$$

where the function *Rank*(x_i) represents the rank of agent x_i. With the probability assigned by Eq. 4.11, it is clear that solutions with poor quality and, therefore, high rank will have a higher probability of changing strategy. According to this mechanism, strategies are refined by eliminating those that produce poor results when compared collectively with other elements of the population. Once Pr_i has been calculated for agent $\mathrm{x_i}$, the next step is to apply the probability law. To do this, a uniformly distributed random value r is calculated. If the value of r is less than the probability Pr_i, the strategy of x_i is modified. Otherwise, no change is necessary.

The strategy of an agent is defined by its position x_i and the mutation rate σ_i, which control the motion of the particle. Therefore, the first element of its adaptation involves changing its position, which means that x_i is modified by the new value x_i^{k+1}. The second element that is modified is the mutation rate σ_i. In the EGT adaptation mechanism, agent x_i is compared to agent x_j. In this relationship, agent x_i is of low quality with respect to agent x_j. Therefore, to improve its strategy with adaptation, agent x_i adopts the mutation rate σ_j of agent x_j as its mutation rate ($\sigma_i = \sigma_j$).

4.4.5 *Flowchart*

Figure 4.5a illustrates the flowchart outlining the operations involved in strategy adaptation using evolutionary strategies. This flowchart includes two connection points (A and B), which link to the segment on strategy adaptation employing the EGT mechanism, to be discussed in the following subsection. The adaptation process conducted by EGT is depicted in Fig. 4.5b. This flowchart highlights the connection points with the strategy adaptation realized by the evolutionary strategies detailed in Sect. 4.4.2. The adaptation strategy for the low-quality individual x_i outlined in the flowchart, follows the steps described in Sect. 4.4.3.

4.5 Experimental Results of ES-EGT

A series of experiments were conducted to assess the effectiveness of the described algorithm. The algorithm's performance results were compared with those obtained by a diverse array of established metaheuristic algorithms. This comparison includes widely recognized algorithms such as Particle Swarm Optimization (PSO) [31], Harmony Search (HS) [32], Bat Algorithm (BA) [33, 34], and State-of-Matter Search

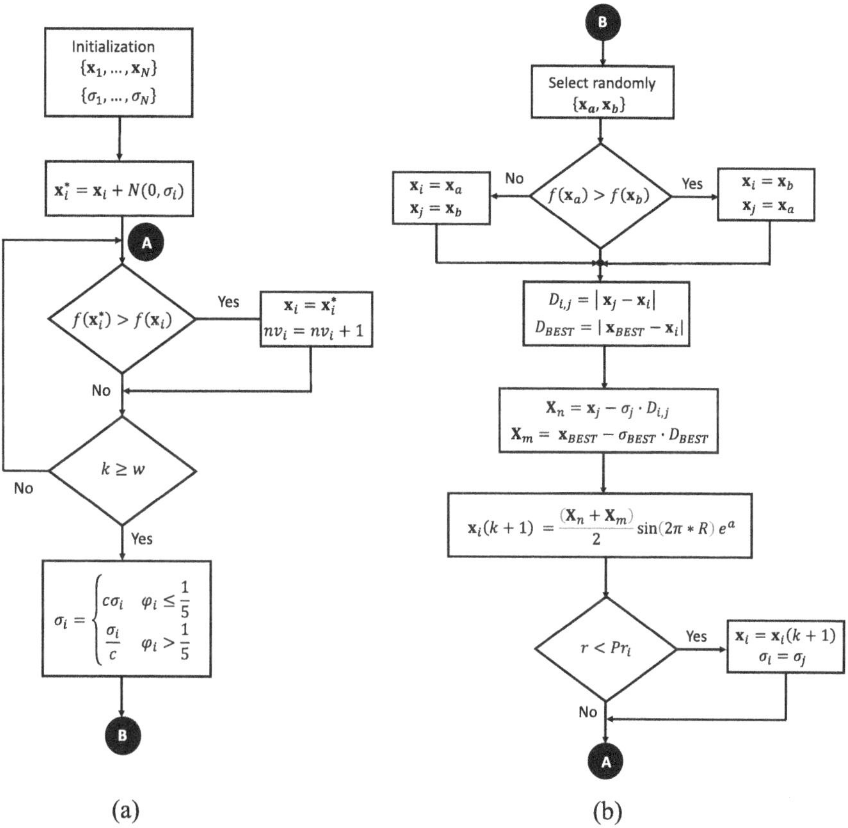

Fig. 4.5 Flow diagrams of the two processes in which is composed the proposed approach: **a** adaptation using ES and **b** adaptation using the EGT mechanism

(SMS) [35]. Additionally, other more recent introductions in this field, such as Moth-Flame Optimization (MFO) [36], Crow Search Algorithm (CSA) [37], and Onlooker Bee Search for Modeling (OBCS) [38], are also included in the comparative analysis. The rationale for this selection is to maintain a balance between popular and well-known algorithms and those that have been recently introduced, providing a comprehensive overview of the proposed algorithm's performance relative to a broad spectrum of existing methods.

In the experiments, each algorithm was configured according to the parameter values suggested by the authors in the foundational papers where these algorithms were first introduced. The experiments aimed to ensure a fair and accurate comparison of the different algorithms' performance by adhering to these prescribed configurations. A predefined upper limit of 50,000 iterations was set as the termination criterion for efficient evaluation of the optimization process. Throughout the experimental assessment, a uniform population size of 100 individuals was maintained across all metaheuristic methods. This chosen population size determines the number of candidate solutions generated and evaluated during each algorithm iteration.

The algorithms were evaluated using a large set of 28 objective functions selected to represent a wide range of complexities. The set of functions includes various formulations, such as unimodal functions, which are relatively more straightforward and have a single global optimum; multimodal functions, which are more complex and have multiple local optima; and composite functions, which combine features of both unimodal and multimodal functions. These functions were examined in several dimensions, namely 30-, 50-, and 100-dimensional spaces, to provide a comprehensive performance evaluation. This approach ensures a multifaceted and detailed assessment of each algorithm's ability to handle different optimization problems, from relatively simple to very complex, at multiple levels of complexity.

In the experiments, each algorithm was run 30 times to mitigate the stochastic nature of its results. This repetition helps minimize random variations in the results, ensuring a more robust and reliable analysis. From these 30 runs, three different statistical indices were used for comparative purposes: Best Average (BA), Best Median (BM), and Standard Deviation (SD). The AB index calculates the average of the best values found in the 30 runs, providing an idea of the typical performance level of the algorithm. The MD index, on the other hand, represents the median of these values, offering an intermediate perspective that is less affected by outliers. Finally, the SD index denotes the standard deviation, measuring the variation or dispersion in the results. Together, these three indices provide a comprehensive view of the algorithms' performance, capturing both the central tendency and the dispersion of the results, allowing for a balanced comparison of their effectiveness.

The experimental results section is methodically segmented into four distinct parts for a comprehensive analysis. The first part evaluates the performance of the proposed algorithm and its counterparts, focusing on their ability to solve the set of 28 functions within a 30-dimensional space. This initial analysis lays the foundation for understanding each algorithm's capabilities in a relatively constrained dimensional environment. The second part extends the dimensionality to 50, examining how these algorithms' performance adjusts when dealing with the same functions in a more complex, higher-dimensional space.

The third part increases the challenge by evaluating the algorithms' performance with the functions extended to 100 dimensions, providing insights into their scalability and effectiveness in even more complex environments. The last part of the results section offers a comparative analysis of the algorithms' convergence rates. It examines how quickly and effectively each algorithm approaches an optimal solution, providing valuable information on their efficiency and effectiveness in different dimensional environments.

4.5.1 Comparison of Results in 30 Dimensions

It is important to note that, with fewer dimensions, the optimization task is typically less complex than scenarios with larger dimensions. The results of this evaluation are summarized in Table 4.1, which details the performance of each algorithm according to three key indices: Best Average (AB), Best Median (MD), and Standard Deviation (SD). The data in Table 4.1 shows that the proposed ES-EGT algorithm outperforms the others in most cases. Specifically, the ES-EGT algorithm achieves superior results in 18 of the 28 tested functions. In 4 functions, the proposed algorithm's performance is on par with at least one other algorithm. However, there are 6 functions where the ES-EGT algorithm does not perform as well as its counterparts. These results highlight the overall effectiveness and capability of the proposed approach in various optimization scenarios.

According to the data presented in Table 4.1, the Onlooker Bee Search for Modeling (OBCS), Moth-Flame Optimization (MFO), and State of Matter Search (SMS) algorithms emerge as notable performers, jointly securing second place in terms of the results they produced. These algorithms demonstrated a commendable level of efficiency, closely following the ES-EGT algorithm, which obtained the best results in handling the optimization tasks on the set of 28 functions. On the other hand, the other algorithms in the study generally showed inferior performance. This distinction in performance levels is significant, as it highlights the relative strengths and weaknesses of different metaheuristic approaches in tackling the same set of optimization problems. While the OBCS, MFO, and SMS algorithms proved to be strong contenders, effectively solving many of the tasks, the other algorithms struggled to match this level of competence, highlighting a clear differentiation in their optimization capabilities within the tested 30-dimensional space.

A significant aspect of the proposed ES-EGT algorithm's performance is its consistency, as evidenced by the standard deviation (SD) index. The algorithm consistently achieves the lowest values in this index, indicating its ability to produce solutions with minimal variability. This robustness, as seen in Table 4.1, highlights the algorithm's reliability and effectiveness in generating consistently high-quality solutions across a variety of functions and scenarios.

The outstanding performance of the proposed ES-EGT algorithm can largely be attributed to its innovative dual adaptation mechanism. This mechanism combines two distinct but complementary approaches. The first part originates from the evolutionary strategy algorithm, which enables each candidate solution to refine and improve its strategy based on its performance continuously. This ensures that each solution evolves, adapts, and improves in a manner tailored to its unique characteristics and performance metrics.

The second crucial component of the dual adaptation mechanism is derived from Evolutionary Game Theory (EGT) principles. In this approach, strategy adaptation is not limited to individual performance but also involves the broader integration of performance across various agents within the population. This collective aspect of strategy adaptation allows for comprehensive and holistic improvement by leveraging

Table 4.1 Minimization results for 28 benchmark functions in 30 dimensions

Function		PSO	HS	BA	SMS	MFO	CSA	OBCS	ES-EGT
$f(x)_1$	AB	1.33E-01	1.23E+01	1.50E+01	2.15E+01	1.05E+01	1.72E+01	1.09E+00	**0.00E+00**
	MD	4.33E-06	1.25E+01	1.49E+01	2.15E+01	1.46E+01	1.73E+01	1.70E-05	**0.00E+00**
	SD	4.09E-01	7.53E-01	1.08E+00	7.68E-02	7.98E+00	3.14E-01	1.86E+00	**0.00E+00**
$f(x)_2$	AB	1.76E+02	1.91E+04	7.32E+01	6.58E+06	4.49E+04	1.56E+05	2.43E+02	**9.93E-01**
	MD	2.08E+02	1.90E+04	1.21E+00	6.53E+06	9.25E+01	1.54E+05	1.32E+02	**1.00E+00**
	SD	1.51E+02	3.84E+03	2.58E+02	5.31E+05	8.05E+04	3.89E+04	3.16E+02	**2.56E-02**
$f(x)_3$	AB	1.19E-02	4.41E+01	1.67E+02	1.25E+03	3.04E+00	1.63E+02	4.92E-01	**0.00E+00**
	MD	9.86E-03	4.34E+01	1.59E+02	1.24E+03	9.90E-03	1.67E+02	2.97E-01	**0.00E+00**
	SD	1.41E-02	8.33E+00	5.28E+01	7.24E+01	1.66E+01	2.11E+01	6.92E-01	**0.00E+00**
$f(x)_4$	AB	**0.00E+00**	1.02E-02	4.87E-17	1.61E+01	1.86E-14	1.21E-01	1.13E-06	**0.00E+00**
	MD	**0.00E+00**	8.90E-03	4.58E-17	1.58E+01	1.42E-16	1.16E-01	1.62E-37	**0.00E+00**
	SD	**0.00E+00**	4.64E-03	2.11E-17	1.62E+00	9.29E-14	3.42E-02	6.14E-06	**0.00E+00**
$f(x)_5$	AB	**1.85E-01**	1.19E+01	1.36E+01	6.15E+02	3.05E+01	5.68E+01	1.09E+00	3.11E+00
	MD	2.44E-10	1.17E+01	1.25E+01	6.14E+02	2.89E+01	5.71E+01	**2.80E-15**	3.17E+00
	SD	3.05E-01	1.52E+00	5.24E+00	3.99E+01	1.11E+01	6.87E+00	2.74E+00	**1.47E-01**
$f(x)_6$	AB	9.47E-31	1.50E-03	2.53E-15	3.50E+01	5.35E-07	9.09E-02	2.80E-05	**2.27E-128**
	MD	**0.00E+00**	1.53E-03	1.68E-15	3.52E+01	**0.00E+00**	9.33E-02	1.51E-05	**0.00E+00**
	SD	5.18E-30	4.14E-04	2.23E-15	2.64E+00	2.93E-06	1.88E-02	3.77E-05	**1.25E-127**
$f(x)_7$	AB	**0.00E+00**	3.02E-38	2.89E+06	7.20E+24	4.66E-197	8.37E+01	6.49E-23	**0.00E+00**
	MD	**0.00E+00**	6.90E-42	1.23E+00	2.70E+24	1.09E-221	1.60E+01	3.78E-38	**0.00E+00**
	SD	**0.00E+00**	1.48E-37	1.50E+07	1.86E+25	**0.00E+00**	1.47E+02	3.54E-22	**0.00E+00**
$f(x)_8$	AB	7.83E+13	8.44E+67	3.28E+66	4.09E+92	6.16E+34	5.33E+75	8.45E+45	**5.80E+03**

(continued)

Table 4.1 (continued)

Function		PSO	HS	BA	SMS	MFO	CSA	OBCS	ES-EGT
	MD	**4.46E+02**	1.38E+67	6.37E+56	3.60E+92	2.02E+23	5.82E+73	2.15E+03	6.06E+03
	SD	4.29E+14	1.79E+68	1.22E+67	1.89E+92	3.36E+35	2.13E+76	4.54E+46	**1.64E+03**
$f(x)_9$	AB	**3.00E+01**	3.46E+01	5.73E+01	1.28E+02	3.43E+01	5.85E+01	3.01E+01	**3.00E+01**
	MD	**3.00E+01**	3.50E+01	5.70E+01	1.28E+02	3.50E+01	5.80E+01	**3.00E+01**	**3.00E+01**
	SD	**0.00E+00**	1.50E+00	5.19E+00	3.59E+00	3.46E+00	1.68E+00	5.71E-01	**0.00E+00**
$f(x)_{10}$	AB	1.09E+01	1.46E+01	2.14E+01	4.46E+02	1.12E+01	2.37E+01	1.14E+01	**9.91E+00**
	MD	1.10E+01	1.45E+01	2.01E+01	4.34E+02	1.03E+01	2.34E+01	1.12E+01	**9.95E+00**
	SD	7.70E-01	7.54E-01	6.72E+00	4.56E+01	2.62E+00	3.68E+00	1.91E+00	**4.98E-01**
$f(x)_{11}$	AB	5.21E+01	9.51E+01	9.12E+01	6.83E+02	1.40E+02	2.63E+02	6.42E+01	**0.00E+00**
	MD	4.73E+01	9.55E+01	9.05E+01	6.75E+02	1.35E+02	2.64E+02	6.53E+01	**0.00E+00**
	SD	2.14E+01	9.96E+00	3.61E+01	2.31E+01	3.16E+01	1.11E+01	1.52E+01	**0.00E+00**
$f(x)_{12}$	AB	2.57E+04	1.60E+04	7.59E+01	6.82E+06	4.70E+04	2.22E+05	**6.69E+01**	3.41E+04
	MD	2.53E+03	1.53E+04	7.45E+01	6.75E+06	1.38E+04	2.20E+05	**3.48E+01**	3.55E+04
	SD	3.14E+04	4.38E+03	**7.78E+01**	6.11E+05	5.69E+04	5.13E+04	9.42E+01	2.80E+04
$f(x)_{13}$	AB	5.96E+00	4.04E+01	4.25E+01	1.00E+02	5.54E+01	5.14E+01	4.30E+00	**1.86E-85**
	MD	5.67E+00	4.10E+01	4.24E+01	1.00E+02	5.51E+01	5.19E+01	6.07E-01	**7.67E-179**
	SD	1.74E+00	2.04E+00	9.90E+00	**0.00E+00**	1.17E+01	3.57E+00	7.28E+00	9.93E-85
$f(x)_{14}$	AB	1.91E+02	1.56E+02	4.23E+29	7.71E+57	5.23E+02	1.57E+30	2.76E+02	**6.48E-83**
	MD	2.15E+02	1.66E+02	2.18E+25	3.77E+51	5.50E+02	2.66E+29	2.52E-01	**1.12E-293**
$f(x)_{15}$	AB	**−9.06E+03**	−1.09E+04	−4.99E+175	2.49E+03	−8.87E+03	−8.23E+03	−9.16E+03	−2.75E+03
	MD	**−9.11E+03**	−1.09E+04	−7.14E+171	2.40E+03	−8.75E+03	−8.11E+03	−9.09E+03	−2.74E+03
	SD	6.05E+02	2.62E+02	6.55E+04	3.19E-04	5.49E-04	4.82E-04	**1.39E-04**	3.34E+02

(continued)

Table 4.1 (continued)

Function		PSO	HS	BA	SMS	MFO	CSA	OBCS	ES-EGT
$f(x)_{16}$	AB	3.33E-02	4.74E+03	1.71E+04	1.36E-01	2.10E-01	3.03E-07	6.18E+00	**0.00E+00**
	MD	**0.00E+00**	4.94E+03	1.63E+04	1.12E-05	4.70E-07	3.29E-10	1.01E-07	**0.00E+00**
	SD	1.83E-01	8.39E+02	4.84E+03	7.05E-01	1.15E+00	9.44E-07	2.33E+01	**0.00E+00**
$f(x)_{17}$	AB	5.63E+04	4.77E+05	1.21E+06	9.88E+00	**9.70E+00**	9.84E+00	9.72E+00	3.00E+01
	MD	9.83E+03	4.69E+05	1.15E+06	9.93E+00	9.77E+00	9.84E+00	**9.72E+00**	3.00E+01
	SD	7.80E+04	8.25E+04	5.32E+05	4.98E-01	4.42E-01	3.45E-01	4.48E-01	**4.94E-02**
$f(x)_{18}$	AB	−2.74E+01	**−2.78E+01**	−2.51E+14	1.78E-04	9.48E-05	1.74E-04	4.68E-06	−5.50E+00
	MD	−2.80E+01	**−2.79E+01**	−2.58E+14	4.20E-05	2.21E-05	1.33E-05	1.67E-06	−5.10E+00
	SD	1.95E+00	2.34E-01	6.09E+13	6.70E-04	1.47E-04	6.18E-04	**8.56E-06**	1.80E+00
$f(x)_{19}$	AB	**0.00E+00**	1.79E+02	5.32E+02	1.46E+02	1.52E+02	1.49E+02	1.22E+02	3.85E+07
	MD	1.68E+02	1.78E+02	3.87E+02	1.51E+02	1.64E+02	1.49E+02	**1.26E+02**	8.44E+06
	SD	9.88E+01	2.52E+01	4.27E+02	2.82E+01	3.00E+01	**2.43E+01**	2.78E+01	5.59E+07
$f(x)_{20}$	AB	4.01E+03	2.51E+04	6.08E+04	9.99E+05	2.23E+04	1.17E+05	9.09E+01	**6.02E-182**
	MD	2.15E+03	2.55E+04	5.53E+04	9.90E+05	1.07E+04	1.19E+05	3.57E-12	**0.00E+00**
	SD	4.91E+03	4.37E+03	3.10E+04	6.85E+04	2.47E+04	1.71E+04	4.97E+02	**0.00E+00**
$f(x)_{21}$	AB	2.27E+05	6.74E+05	3.43E+06	4.90E+07	6.13E+05	4.19E+06	4.81E+04	**7.81E-190**
	MD	1.40E+05	6.68E+05	3.22E+06	4.91E+07	2.80E+05	4.26E+06	4.29E+04	**0.00E+00**
	SD	2.35E+05	1.06E+05	1.47E+06	1.86E+06	8.13E+05	5.64E+05	3.83E+04	**0.00E+00**
$f(x)_{22}$	AB	4.35E-13	1.24E+01	1.32E-05	3.72E+02	8.74E-01	5.36E+01	8.47E-02	**1.22E-139**
	MD	9.87E-14	1.29E+01	1.34E-05	3.68E+02	9.21E-08	5.41E+01	5.60E-08	**0.00E+00**
	SD	8.02E-13	2.01E+00	1.68E-06	1.52E+01	4.79E+00	5.89E+00	2.59E-01	**6.71E-139**
$f(x)_{23}$	AB	5.33E+01	5.66E+02	3.07E+01	2.26E+04	3.87E+02	2.43E+03	1.07E+01	**1.08E-151**

(continued)

Table 4.1 (continued)

Function		PSO	HS	BA	SMS	MFO	CSA	OBCS	ES-EGT
	MD	2.93E-11	5.57E+02	5.98E-04	2.24E+04	3.00E+02	2.43E+03	3.99E-01	**0.00E+00**
	SD	8.19E+01	8.24E+01	1.13E+02	1.14E+03	4.66E+02	4.10E+02	2.28E+01	**5.92E-151**
$f(x)_{24}$	AB	7.17E-26	3.36E-05	9.77E-09	2.06E+01	2.20E-21	7.20E-03	**0.00E+00**	**0.00E+00**
	MD	1.10E-29	3.14E-05	9.39E-09	2.10E+01	2.74E-22	6.45E-03	**0.00E+00**	**0.00E+00**
	SD	3.26E-25	1.53E-05	3.48E-09	1.38E+00	7.18E-21	3.83E-03	**0.00E+00**	**0.00E+00**
$f(x)_{25}$	AB	6.69E+02	4.59E+03	2.54E+04	1.76E+22	3.05E+04	2.08E+04	3.63E+01	**1.23E-65**
	MD	7.67E-07	4.63E+03	2.24E+04	6.62E+21	3.01E+04	2.10E+04	4.16E-07	**1.70E-163**
	SD	2.55E+03	6.10E+02	1.81E+04	4.10E+22	2.08E+04	2.50E+03	1.48E+02	**6.74E-65**
$f(x)_{26}$	AB	6.52E+01	3.32E+02	5.82E+02	4.68E+03	1.55E+02	7.48E+02	6.51E+01	**2.90E+01**
	MD	6.22E+01	3.37E+02	5.90E+02	4.68E+03	1.02E+02	7.70E+02	**2.90E+01**	2.90E+01
	SD	1.67E+01	2.57E+01	1.53E+02	3.13E+02	1.05E+02	7.56E+01	5.22E+01	**3.64E-05**
$f(x)_{27}$	AB	9.56E+01	6.22E+06	2.42E+07	3.79E+09	7.41E+01	6.91E+07	3.81E+02	**3.20E+01**
	MD	9.69E+01	6.06E+06	9.85E+06	3.80E+09	6.92E+01	7.06E+07	1.09E+02	**3.20E+01**
	SD	2.26E+01	1.96E+06	3.84E+07	2.65E+08	2.00E+01	1.96E+07	5.28E+02	**0.00E+00**
$f(x)_{28}$	AB	1.15E+02	3.41E+02	1.60E+03	6.82E+22	1.03E+03	9.16E+02	6.99E+01	**2.90E+01**
	MD	6.42E+01	3.48E+02	1.02E+03	1.36E+22	9.34E+02	9.20E+02	2.90E+01	**2.90E+01**
	SD	1.41E+02	3.24E+01	2.61E+03	2.24E+23	5.54E+02	1.22E+02	7.08E+01	**0.00E+00**

the collective experiences and successes of the entire population, resulting in a richer and more diverse search strategy that combines the strengths and expertise of multiple agents.

Together, these two components of the ES-EGT algorithm—individual performance-based refinement and collective strategy adaptation—create a powerful synergy. This dual approach allows the algorithm to be highly responsive to individual success while adapting to broader population trends and dynamics. The result is a robust, flexible, and highly efficient optimization process capable of navigating complex problem spaces, as demonstrated by the algorithm's superior performance in experimental evaluations.

4.5.2 Wilcoxon's Test Analysis on 30 Dimensions

The Wilcoxon signed-rank test [39] is a statistical analysis method frequently used in evaluating metaheuristic algorithms. This test determines the statistical significance of performance differences between algorithms and compares their effectiveness with each other or with a standard benchmark method. Metaheuristic algorithms, being stochastic optimization methods, inherently use randomness in their search for optimal solutions in complex environments. Given this stochastic nature, their performance can vary between different iterations or data sets. The Wilcoxon test is essential to determine whether variations in algorithm performance are substantial or merely incidental. As a nonparametric test, it does not rely on assumptions about the data distribution. It focuses on comparing ranges of paired results, typically those obtained by two different algorithms on identical reference functions. This comparison of ranges leads to calculating a p-value, which quantifies the probability that the observed performance differences occurred by chance. A p-value below a set threshold (e.g., 0.05) indicates that the performance disparity is statistically significant. This result would suggest that one algorithm has a clear performance advantage or disadvantage over the other beyond what would be expected from random variability alone.

Table 4.2 of the study presents the p-values obtained from the Wilcoxon test, offering a statistical perspective based on the performance data in Table 4.1 (where $d = 30$). This table allows for a detailed pairwise comparison of the ES-EGT algorithm with all other tested algorithms, helping to validate the experimental results statistically. To facilitate the visualization and interpretation of these results, Table 4.2 employs specific symbols: ▲, ▼, and ►. The symbol ▲ indicates that the ES-EGT algorithm achieves significantly better results than a competing algorithm. Conversely, the symbol ▼ signifies that the ES-EGT algorithm's results are inferior to its competitors. Lastly, the symbol ► is used when the Wilcoxon test shows no discernible difference between the numerical results of the compared algorithms. This symbolic representation provides a clear and effective way to quickly understand and evaluate the comparative performance levels demonstrated by the ES-EGT algorithm in the experimental analysis.

Table 4.2 Results of Wilcoxon analysis for 28 test functions in 30 dimensions

	ES-EGT VS PSO	ES-EGT VS HS	ES-EGT VS BA	ES-EGT VS SMS	ES-EGT VS MFO	ES-EGT VS CSA	ES-EGT VS OBSC
$f(x)_1$	1.21E-12▲	1.21E-12▲	1.21E-12▲	1.21E-12▲	1.21E-12▲	1.21E-12▲	1.21E-12▲
$f(x)_2$	1.89E-03▲	2.63E-11▲	6.40E-01▲	2.63E-11▲	1.03E-06▲	2.63E-11▲	7.64E-09▲
$f(x)_3$	1.21E-12▲	1.21E-12▲	1.21E-12▲	1.21E-12▲	1.21E-12▲	1.21E-12▲	5.85E-09▲
$f(x)_4$	1.00E+00►	1.21E-12▲	1.21E-12▲	1.21E-12▲	1.21E-12▲	1.21E-12▲	1.70E-08▲
$f(x)_5$	3.02E-11▼	3.02E-11▲	3.02E-11▲	3.02E-11▲	3.02E-11▲	3.02E-11▲	9.50E-06▼
$f(x)_6$	2.31E-03▲	1.44E-11▲	1.44E-11▲	1.42E-11▲	2.31E-03▲	1.44E-11▲	1.44E-11▲
$f(x)_7$	1.00E+00►	1.21E-12▲	1.21E-12▲	1.21E-12▲	1.21E-12▲	1.21E-12▲	1.66E-11▲
$f(x)_8$	6.91E-04▲	3.02E-11▲	1.07E-07▲	3.02E-11▲	3.02E-11▲	3.02E-11▲	8.56E-04▲
$f(x)_9$	1.00E+00►	9.92E-13▲	1.17E-12▲	1.16E-12▲	3.41E-08▲	1.07E-12▲	1.61E-01►
$f(x)_{10}$	2.68E-06▲	3.02E-11▲	2.37E-10▲	3.02E-11▲	9.03E-04▲	3.02E-11▲	1.11E-03▲
$f(x)_{11}$	1.21E-12▲	1.21E-12▲	1.21E-12▲	1.21E-12▲	1.21E-12▲	1.21E-12▲	1.21E-12▲
$f(x)_{12}$	9.05E-02▼	2.24E-02▼	3.34E-11▼	3.02E-11▲	5.01E-01▲	3.02E-11▲	3.69E-11▼
$f(x)_{13}$	3.00E-11▲	3.00E-11▲	3.00E-11▲	1.20E-12▲	3.00E-11▲	3.00E-11▲	4.57E-10▲
$f(x)_{14}$	2.72E-11▲	2.72E-11▲	2.72E-11▲	2.72E-11▲	2.72E-11▲	2.72E-11▲	2.62E-08▲
$f(x)_{15}$	3.02E-11▼	3.02E-11▼	3.02E-11▼	3.02E-11▲	3.02E-11▼	3.02E-11▼	3.02E-11▼
$f(x)_{16}$	3.34E-01▲	1.21E-12▲	1.21E-12▲	1.21E-12▲	1.10E-02▲	1.21E-12▲	4.19E-02▲
$f(x)_{17}$	1.11E-06▲	3.02E-11▲	3.02E-11▲	3.02E-11▲	5.57E-10▲	3.02E-11▲	8.48E-09▲
$f(x)_{18}$	3.02E-11▼	3.02E-11▼	5.88E-08▼	3.02E-11▲	3.02E-11▼	1.62E-11▼	3.02E-11▼
$f(x)_{19}$	3.02E-11▼	3.02E-11▼	8.15E-11▼	3.02E-11▲	3.02E-11▼	3.34E-11▼	3.02E-11▼
$f(x)_{20}$	5.22E-12▲	5.22E-12▲	5.22E-12▲	5.22E-12▲	5.22E-12▲	5.22E-12▲	4.13E-08▲
$f(x)_{21}$	1.10E-11▲	1.10E-11▲	1.10E-11▲	1.10E-11▲	1.10E-11▲	1.10E-11▲	1.10E-11▲
$f(x)_{22}$	1.44E-11▲	1.44E-11▲	1.44E-11▲	1.44E-11▲	1.44E-11▲	1.44E-11▲	2.23E-08▲
$f(x)_{23}$	1.62E-11▲	1.62E-11▲	1.62E-11▲	1.62E-11▲	1.62E-11▲	1.62E-11▲	1.62E-11▲
$f(x)_{24}$	1.21E-12▲	1.21E-12▲	1.21E-12▲	8.84E-13▲	1.21E-12▲	1.21E-12▲	1.00E+00►
$f(x)_{25}$	2.95E-11▲	2.95E-11▲	2.95E-11▲	2.95E-11▲	2.95E-11▲	2.95E-11▲	1.08E-04▲
$f(x)_{26}$	3.02E-11▲	3.02E-11▲	3.02E-11▲	3.02E-11▲	3.02E-11▲	3.02E-11▲	1.00E+00▲
$f(x)_{27}$	1.21E-12▲	1.21E-12▲	1.21E-12▲	1.21E-12▲	1.21E-12▲	1.21E-12▲	1.21E-12▲
$f(x)_{28}$	1.21E-12▲	1.21E-12▲	1.21E-12▲	1.21E-12▲	1.21E-12▲	1.21E-12▲	5.77E-11▲
▲	23	24	24	28	25	25	21
▼	2	4	4	0	3	3	5
►	3	0	0	0	0	0	2

The p-value analysis in Table 4.2 offers an in-depth statistical comparison of the proposed ES-EGT algorithm against other methods. Compared to the Particle Swarm Optimization (PSO) method, ES-EGT shows statistically superior results in 23 functions, inferior results in 2 functions, and no significant difference in 2 functions. Compared with the Harmony Search (HS) method, ES-EGT is statistically superior in 24 functions but lags in 4 functions. Against the Bat Algorithm (BA), ES-EGT excels in 24 functions but is less effective in 4. A remarkable outcome is observed when comparing ES-EGT with the State of Matter Search (SMS) method, where ES-EGT demonstrates superior performance across all 28 functions. Compared to the Moth-flame Optimization (MFO) method, ES-EGT performs better in 25 functions and is less efficient in 3. A similar pattern emerges in comparison with the Crow Search Algorithm (CSA), where ES-EGT shows better results in 25 functions and worse in 3. Finally, compared to the Onlooker Bee Search for Modeling (OBSC) method, the ES-EGT algorithm performs better in 21 functions and worse in 5, with 2 functions showing statistically inconclusive results. These comprehensive comparisons demonstrate the robustness and effectiveness of the ES-EGT algorithm over a wide range of functions and against several established methods.

4.5.3 Comparison of Results on 50 Dimensions

Increasing the dimensions from 30 to 50 naturally raises the complexity of finding the optimal solution, posing a more challenging test for each algorithm. The results of this evaluation are compiled in Table 4.3, which shows the outcomes for the 50-dimensional case using three indices: Best Average (AB), Best Median (MD), and Standard Deviation (SD). The data in Table 4.3 indicate that the proposed ES-EGT algorithm consistently outperforms the others. Specifically, the ES-EGT algorithm achieves better solutions in 20 of the 28 functions tested. Its performance is on par with at least one other algorithm in 2 functions, while in 6 functions, it falls behind its competitors. These results highlight the strengths of the proposed ES-EGT approach, particularly when dealing with the higher complexity of higher-dimensional spaces. A notable observation from Table 4.3 is the superior performance of the ES-EGT algorithm relative to other methods, especially as the dimensions increase. The algorithm maintains its efficiency and improves its comparative performance, as evidenced by the greater number of functions where it surpasses other algorithms. This trend suggests that the ES-EGT algorithm is particularly adept at scaling its problem-solving capabilities in more complex, higher-dimensional environments, further underscoring its robustness and adaptability as a metaheuristic solution.

Table 4.3 also shows that in the 50-dimensional environment, the Moth-Flame Optimization (MFO) and Particle Swarm Optimization (PSO) algorithms rank second in terms of performance. This notable position is significant given the higher complexity presented by the higher-dimensional space. Both MFO and PSO demonstrate a commendable level of efficiency, proving their ability to adapt and solve more complex problems effectively. This achievement distinguishes them from the other tested algorithms, which generally perform worse in this scenario. The less favorable results of the other algorithms highlight the challenges posed by the 50-dimensional space, where only a few methods, like MFO and PSO, maintain high levels of efficiency. Their success in achieving second place reflects the robustness of their design and their suitability for tackling complex optimization tasks, making them stand out in a field where many others struggled to adapt to the increased complexity.

Table 4.3 Minimization results for 28 benchmark functions in 50 dimensions

Function		PSO	HS	BA	SMS	MFO	CSA	OBCS	ES-EGT
$f(x)_1$	AB	2.40E-01	1.57E+01	1.56E+01	2.14E+01	1.58E+01	1.80E+01	8.45E+00	**0.00E+00**
	MD	1.49E-03	1.57E+01	1.58E+01	2.14E+01	1.84E+01	1.81E+01	3.61E+00	**0.00E+00**
	SD	5.51E-01	3.14E-01	1.02E+00	4.41E-02	6.04E+00	2.96E-01	8.71E+00	**0.00E+00**
$f(x)_2$	AB	2.50E+03	2.88E+05	**9.54E-01**	1.63E+07	1.29E+05	7.99E+05	1.97E+04	1.00E+00
	MD	4.67E+01	2.87E+05	**6.73E-01**	1.63E+07	3.26E+02	8.15E+05	5.12E+03	1.00E+00
	SD	1.32E+04	4.12E+04	1.39E+00	1.28E+06	2.61E+05	1.57E+05	3.43E+04	**5.16E-05**
$f(x)_3$	AB	6.16E-03	1.83E+02	2.99E+02	2.00E+03	6.02E+01	3.82E+02	5.12E+00	**0.00E+00**
	MD	3.70E-03	1.84E+02	2.71E+02	2.02E+03	5.05E-02	3.83E+02	4.69E-01	**0.00E+00**
	SD	7.53E-03	1.66E+01	1.02E+02	1.06E+02	8.32E+01	2.74E+01	1.43E+01	**0.00E+00**
$f(x)_4$	AB	**0.00E+00**	1.79E-01	5.57E-16	2.32E+01	7.72E-02	5.27E-01	4.04E-04	**0.00E+00**
	MD	**0.00E+00**	1.76E-01	5.23E-16	2.28E+01	5.76E-12	5.20E-01	7.36E-28	**0.00E+00**
	SD	**0.00E+00**	3.90E-02	2.13E-16	2.36E+00	2.94E-01	1.12E-01	1.69E-03	**0.00E+00**
$f(x)_5$	AB	4.93E+00	5.68E+01	2.30E+01	9.50E+02	6.68E+01	1.36E+02	6.92E+00	**4.92E+00**
	MD	3.50E+00	5.64E+01	2.11E+01	9.44E+02	5.70E+01	1.35E+02	**1.10E-10**	4.97E+00
	SD	5.24E+00	4.96E+00	8.19E+00	6.04E+01	2.95E+01	1.23E+01	1.39E+01	**1.44E-01**
$f(x)_6$	AB	**0.00E+00**	1.01E-02	8.37E-15	3.41E+01	1.13E-05	1.85E-01	1.46E-04	1.25E-182
	MD	**0.00E+00**	1.03E-02	7.20E-15	3.40E+01	**0.00E+00**	1.84E-01	4.26E-11	**0.00E+00**
	SD	**0.00E+00**	1.77E-03	4.60E-15	2.56E+00	6.21E-05	2.85E-02	4.89E-04	**0.00E+00**
$f(x)_7$	AB	**0.00E+00**	1.06E-40	2.79E+12	5.08E+38	**0.00E+00**	2.67E+06	5.13E-26	**0.00E+00**
	MD	**0.00E+00**	3.14E-47	1.45E+02	6.20E+36	**0.00E+00**	5.44E+05	6.74E-45	**0.00E+00**
	SD	**0.00E+00**	5.75E-40	1.48E+13	1.86E+39	**0.00E+00**	5.85E+06	2.81E-25	**0.00E+00**
$f(x)_8$	AB	1.09E+107	1.24E+150	1.88E+136	1.44E+174	2.25E+126	1.04E+154	4.54E+144	**1.07E+04**

(continued)

Table 4.3 (continued)

Function		PSO	HS	BA	SMS	MFO	CSA	OBCS	ES-EGT
	MD	1.23E+93	1.28E+149	1.15E+127	1.35E+174	1.13E+116	1.33E+151	2.98E+36	**1.06E+04**
	SD	5.95E+107	2.32E+150	1.03E+137	6.55E+04	1.04E+127	6.55E+04	2.49E+145	**1.92E+03**
$f(x)_9$	AB	3.16E+01	5.21E+01	7.34E+01	1.85E+02	3.93E+01	8.53E+01	3.23E+01	**3.01E+01**
	MD	**3.00E+01**	5.30E+01	7.40E+01	1.83E+02	4.00E+01	8.60E+01	**3.00E+01**	**3.00E+01**
	SD	2.57E+00	2.01E+00	6.55E+00	5.64E+00	6.26E+00	3.35E+00	4.89E+00	**2.54E-01**
$f(x)_{10}$	AB	2.25E+01	4.44E+01	4.87E+01	1.08E+03	4.62E+01	7.68E+01	2.58E+01	**1.83E+01**
	MD	2.18E+01	4.52E+01	4.45E+01	1.08E+03	3.36E+01	7.70E+01	2.47E+01	**1.84E+01**
	SD	2.30E+00	3.80E+00	1.47E+01	6.50E+01	2.80E+01	1.13E+01	7.12E+00	**7.40E-01**
$f(x)_{11}$	AB	1.71E+02	2.63E+02	1.31E+02	1.08E+03	2.70E+02	4.93E+02	1.85E+02	**0.00E+00**
	MD	1.72E+02	2.63E+02	1.25E+02	1.08E+03	2.56E+02	4.90E+02	1.97E+02	**0.00E+00**
	SD	3.23E+01	1.25E+01	3.75E+01	3.72E+01	6.47E+01	1.54E+01	4.86E+01	**0.00E+00**
$f(x)_{12}$	AB	2.54E+05	1.37E+05	1.16E+02	1.02E+07	2.54E+05	7.01E+05	**5.37E+02**	4.85E+04
	MD	2.24E+05	1.42E+05	8.82E+01	1.02E+07	2.97E+05	7.19E+05	**7.68E+01**	4.78E+04
	SD	1.32E+05	2.35E+04	**1.57E+02**	6.68E+05	1.30E+05	1.18E+05	1.55E+03	3.24E+04
$f(x)_{13}$	AB	2.51E+01	5.85E+01	4.92E+01	1.00E+02	7.57E+01	6.05E+01	1.74E+01	**4.67E-105**
	MD	2.53E+01	5.89E+01	4.97E+01	1.00E+02	7.80E+01	6.04E+01	6.64E+00	**0.00E+00**
	SD	2.82E+00	1.79E+00	8.75E+00	**0.00E+00**	6.88E+00	2.54E+00	2.23E+01	2.56E-104
$f(x)_{14}$	AB	5.52E+02	3.66E+06	1.51E+56	8.53E+85	9.37E+02	3.30E+54	4.39E+21	**1.02E-192**
	MD	5.45E+02	1.38E+04	1.65E+50	6.92E+84	1.00E+03	5.55E+53	9.74E+03	**0.00E+00**
	SD	1.93E+02	1.37E+07	6.58E+56	2.29E+86	3.21E+02	7.47E+54	1.88E+22	**0.00E+00**
$f(x)_{15}$	AB	−1.39E+04	**−1.55E+04**	−4.65E+277	3.38E+03	−1.36E+04	−1.12E+04	−1.41E+04	−3.51E+03
	MD	−1.41E+04	**−1.55E+04**	−4.20E+272	3.29E+03	−1.39E+04	−1.09E+04	−1.37E+04	−3.50E+03

(continued)

Table 4.3 (continued)

Function		PSO	HS	BA	SMS	MFO	CSA	OBCS	ES-EGT
	SD	1.01E+03	4.07E+02	6.55E+04	6.60E+02	1.65E+03	1.13E+03	9.55E+02	**4.04E+02**
$f(x)_{16}$	AB	7.33E-01	1.97E+04	3.10E+04	2.23E+05	6.01E+03	4.18E+04	3.76E+02	**2.33E-01**
	MD	1.00E+00	1.97E+04	3.00E+04	2.22E+05	6.95E+01	4.23E+04	**0.00E+00**	**0.00E+00**
	SD	7.85E-01	2.06E+03	8.61E+03	1.08E+04	7.70E+03	3.09E+03	1.45E+03	**6.79E-01**
$f(x)_{17}$	AB	1.10E+06	1.41E+07	1.83E+07	1.59E+08	3.26E+06	2.15E+07	1.19E+05	**4.99E+01**
	MD	9.10E+05	1.40E+07	1.91E+07	1.56E+08	2.57E+06	2.18E+07	4.10E+02	**5.00E+01**
	SD	9.06E+05	1.30E+06	4.78E+06	1.13E+07	2.87E+06	2.31E+06	2.48E+05	**1.10E-01**
$f(x)_{18}$	AB	−4.14E+01	−4.23E+01	−3.87E+14	1.09E+01	**−4.75E+01**	−9.80E+13	−4.57E+01	−6.22E+00
	MD	−4.24E+01	−4.25E+01	−3.94E+14	1.07E+01	**−4.81E+01**	−1.01E+14	−4.60E+01	−6.02E+00
	SD	3.19E+00	**6.43E-01**	1.04E+14	2.28E+00	1.74E+00	2.26E+13	1.12E+00	3.12E+00
$f(x)_{19}$	AB	8.81E+02	4.71E+02	1.05E+03	5.59E+13	6.93E+02	2.06E+03	**3.86E+02**	5.20E+08
	MD	8.69E+02	4.83E+02	8.95E+02	5.34E+13	6.88E+02	1.86E+03	**2.38E+02**	1.07E+08
	SD	2.59E+02	**4.75E+01**	8.36E+02	1.42E+13	1.42E+02	1.01E+03	3.56E+02	1.02E+09
$f(x)_{20}$	AB	2.06E+04	1.66E+05	1.08E+05	2.46E+06	9.35E+04	4.15E+05	3.78E+02	**7.70E-288**
	MD	1.07E+04	1.70E+05	1.07E+05	2.43E+06	8.16E+04	4.21E+05	1.09E-07	**0.00E+00**
	SD	2.50E+04	1.87E+04	8.77E+04	1.15E+05	7.59E+04	4.42E+04	1.39E+03	**0.00E+00**
$f(x)_{21}$	AB	1.89E+06	9.58E+06	1.34E+07	2.03E+08	9.73E+06	2.75E+07	2.26E+06	**9.51E-207**
	MD	1.90E+06	9.49E+06	1.42E+07	2.00E+08	7.96E+06	2.82E+07	1.52E+06	**0.00E+00**
	SD	1.46E+06	8.05E+05	6.90E+06	1.39E+07	7.24E+06	2.73E+06	2.23E+06	**0.00E+00**
$f(x)_{22}$	AB	1.68E-09	5.37E+01	3.85E-05	5.81E+02	1.84E+01	1.13E+02	4.59E-01	**1.27E-230**
	MD	2.43E-10	5.43E+01	3.86E-05	5.73E+02	2.62E+01	1.13E+02	8.04E-07	**0.00E+00**
	SD	6.88E-09	4.30E+00	3.94E-06	2.57E+01	2.30E+01	8.93E+00	1.54E+00	**0.00E+00**

(continued)

Table 4.3 (continued)

Function		PSO	HS	BA	SMS	MFO	CSA	OBCS	ES-EGT
$f(x)_{23}$	AB	5.07E+02	3.95E+03	1.91E+00	5.97E+04	1.73E+03	9.12E+03	7.55E+02	**3.86E-245**
	MD	5.00E+02	3.94E+03	3.50E-03	5.97E+04	1.10E+03	9.12E+03	2.07E+02	**0.00E+00**
	SD	3.89E+02	4.32E+02	7.52E+00	2.93E+03	1.84E+03	1.04E+03	1.22E+03	**0.00E+00**
$f(x)_{24}$	AB	2.19E-19	1.43E-04	4.52E-09	3.26E+01	7.41E-12	9.72E-03	4.43E-16	**1.29E-297**
	MD	5.17E-23	1.27E-04	4.11E-09	3.20E+01	4.03E-19	7.55E-03	**0.00E+00**	**0.00E+00**
	SD	1.18E-18	6.70E-05	1.54E-09	1.43E+00	4.06E-11	6.95E-03	2.39E-15	**0.00E+00**
$f(x)_{25}$	AB	8.03E+03	1.79E+04	3.45E+06	1.13E+36	5.22E+04	9.19E+04	8.71E+02	**1.67E-157**
	MD	1.00E+04	1.82E+04	4.38E+04	2.38E+35	5.52E+04	6.80E+04	5.41E-04	**0.00E+00**
	SD	8.50E+03	2.34E+03	1.42E+07	2.85E+36	3.06E+04	6.63E+04	2.60E+03	**9.15E-157**
$f(x)_{26}$	AB	1.47E+02	9.60E+02	8.77E+02	7.04E+03	4.41E+02	1.50E+03	2.96E+02	**4.90E+01**
	MD	1.49E+02	9.67E+02	8.44E+02	7.03E+03	4.61E+02	1.50E+03	3.78E+02	**4.90E+01**
	SD	1.99E+01	4.33E+01	1.92E+02	3.45E+02	2.49E+02	6.94E+01	1.85E+02	**5.08E-05**
$f(x)_{27}$	AB	5.71E+02	8.46E+07	4.38E+07	5.61E+09	3.45E+02	2.34E+08	8.31E+04	**5.40E+01**
	MD	5.59E+02	8.57E+07	2.83E+07	5.64E+09	3.23E+02	2.43E+08	2.87E+02	**5.40E+01**
	SD	1.26E+02	1.63E+07	4.64E+07	3.25E+08	9.59E+01	5.42E+07	2.76E+05	**3.64E-07**
$f(x)_{28}$	AB	5.60E+02	9.27E+02	9.70E+04	2.47E+36	1.93E+03	3.95E+04	2.77E+02	**4.90E+01**
	MD	5.05E+02	9.33E+02	1.48E+03	2.71E+35	1.68E+03	4.55E+03	7.77E+01	**4.90E+01**
	SD	3.58E+02	7.16E+01	5.08E+05	7.47E+36	1.02E+03	1.12E+05	2.64E+02	**0.00E+00**

The impressive performance of the proposed algorithm is attributed to its innovative adaptation methods, which play a crucial role in enhancing the quality of its strategies during the exploration of the search space. These methods allow the algorithm to dynamically and intelligently adjust its approach in response to the evolving landscape of the problem it is solving. As the algorithm navigates the search space, it continually evaluates and refines its strategies, learning and adapting effectively based on the results it encounters. This self-improving capability is especially advantageous in complex optimization tasks, where the search space is vast and full of potential pitfalls, such as local optima. The adaptation methods employed are designed to balance exploration with exploitation: the algorithm is adept at discovering promising new areas of the search space (exploration) while effectively refining and capitalizing on the successful strategies it has already identified (exploitation). This dynamic adaptability ensures that the algorithm remains effective, even as the complexity of the problem increases, whether due to size, dimensionality, or the inherent complexities of the search space. In essence, the robust performance of the proposed algorithm is a direct result of its ability to adapt and evolve its strategies in real-time. This allows it to continuously optimize its approach, resulting in high-quality solutions in a variety of challenging optimization scenarios.

4.5.4 Wilcoxon's Test Analysis on 50 Dimensions

Table 4.4 of the study provides an in-depth statistical analysis using p-values derived from the Wilcoxon test based on the performance data compiled in Table 4.3 for the 50-dimensional case ($d = 50$). This table is specifically designed to offer a comprehensive pairwise comparison between the proposed ES-EGT algorithm and each of the other algorithms tested. The inclusion of these p-values is vital, as it adds statistical rigor to the evaluation process. By employing the Wilcoxon test, Table 4.4 offers a nuanced understanding of the statistical standing of the ES-EGT algorithm relative to its counterparts. This method of analysis is particularly valuable for validating experimental results, as it provides a clear picture of the relative performance and efficiency of the studied algorithms. It ensures that conclusions drawn from experimental results are not based solely on observed trends but are also supported by statistically significant evidence, which strengthens the validity and reliability of the results.

The p-value analysis in Table 4.4 presents significant statistical evidence of the ES-EGT algorithm's performance compared to other methods in a 50-dimensional setting. When compared to the PSO method, ES-EGT demonstrates statistically superior results on 20 features but is outperformed on 4, with 4 features showing no clear statistical winner. In comparison with the HS method, ES-EGT excels in 25 functions and falls short in 3. Compared to the BA method, ES-EGT again stands out with better results in 23 functions, inferior results in 4, and no significant difference in 1 function. A notable achievement is seen in the comparison with the SMS method, where ES-EGT shows superior statistical results in all 28 functions. When

Table 4.4 Results of Wilcoxon analysis for 28 test functions in 50 dimensions

ES-EGT VS PSO		ES-EGT VS HS	ES-EGT VS BA	ES-EGT VS SMS	ES-EGT VS MFO	ES-EGT VS CSA	ES-EGT VS OBSC
$f(x)_1$	1.21E-12▲	1.21E-12▲	1.21E-12▲	1.21E-12▲	1.21E-12▲	1.21E-12▲	1.21E-12▲
$f(x)_2$	8.32E-07▲	1.79E-11▲	3.53E-10►	1.79E-11▲	1.79E-11▲	1.79E-11▲	1.79E-11▲
$f(x)_3$	1.21E-12▲	1.21E-12▲	1.21E-12▲	1.21E-12▲	1.21E-12▲	1.21E-12▲	1.95E-09▲
$f(x)_4$	1.00E+00►	1.21E-12▲	1.21E-12▲	1.21E-12▲	1.21E-12▲	1.21E-12▲	1.93E-10▲
$f(x)_5$	8.77E-02►	3.02E-11▲	3.02E-11▲	3.02E-11▲	3.02E-11▲	3.02E-11▲	7.96E-03▲
$f(x)_6$	4.19E-02▼	4.11E-12▲	4.11E-12▲	4.10E-12▲	4.69E-01▲	4.11E-12▲	4.11E-12▲
$f(x)_7$	1.00E+00►	1.21E-12▲	1.21E-12▲	1.21E-12▲	1.00E+00►	1.21E-12▲	1.66E-11▲
$f(x)_8$	3.02E-11▲	3.02E-11▲	3.02E-11▲	3.02E-11▲	3.02E-11▲	3.02E-11▲	1.27E-02▲
$f(x)_9$	1.29E-02►	2.03E-12▲	2.25E-12▲	2.29E-12▲	3.94E-11▲	2.16E-12▲	1.01E-01▲
$f(x)_{10}$	3.02E-11▲	3.02E-11▲	3.02E-11▲	3.02E-11▲	3.02E-11▲	3.02E-11▲	7.74E-06▲
$f(x)_{11}$	1.21E-12▲	1.21E-12▲	1.21E-12▲	1.21E-12▲	1.21E-12▲	1.21E-12▲	1.21E-12▲
$f(x)_{12}$	1.55E-09▲	1.33E-10▲	3.69E-11▼	3.02E-11▲	4.57E-09▲	3.02E-11▲	8.15E-11▼
$f(x)_{13}$	1.96E-11▲	1.96E-11▲	1.96E-11▲	6.87E-13▲	1.96E-11▲	1.96E-11▲	1.96E-11▲
$f(x)_{14}$	2.26E-11▲	2.26E-11▲	2.26E-11▲	2.26E-11▲	2.26E-11▲	2.26E-11▲	2.26E-11▲
$f(x)_{15}$	3.02E-11▼	3.02E-11▼	3.02E-11▼	3.02E-11▲	3.02E-11▼	3.02E-11▼	3.02E-11▼
$f(x)_{16}$	1.20E-03▲	4.11E-12▲	4.11E-12▲	4.11E-12▲	3.97E-07▲	4.11E-12▲	1.10E-01▲
$f(x)_{17}$	5.57E-10▲	3.02E-11▲	3.02E-11▲	3.02E-11▲	3.02E-11▲	3.02E-11▲	1.00E+00▲
$f(x)_{18}$	3.02E-11▼	3.02E-11▼	1.00E+00▼	3.02E-11▲	3.02E-11▼	4.39E-12▼	3.02E-11▼
$f(x)_{19}$	3.02E-11▼	3.02E-11▼	3.02E-11▼	3.02E-11▲	3.02E-11▼	3.02E-11▼	3.02E-11▼
$f(x)_{20}$	1.72E-12▲	1.72E-12▲	1.72E-12▲	1.72E-12▲	1.72E-12▲	1.72E-12▲	3.51E-10▲
$f(x)_{21}$	6.48E-12▲	6.48E-12▲	6.48E-12▲	6.48E-12▲	6.48E-12▲	6.48E-12▲	6.48E-12▲
$f(x)_{22}$	9.40E-12▲	9.40E-12▲	9.40E-12▲	9.39E-12▲	9.40E-12▲	9.40E-12▲	5.69E-11▲
$f(x)_{23}$	2.37E-12▲	2.37E-12▲	2.37E-12▲	2.37E-12▲	2.37E-12▲	2.37E-12▲	2.37E-12▲
$f(x)_{24}$	1.72E-12▲	1.72E-12▲	1.72E-12▲	1.42E-12▲	1.72E-12▲	1.72E-12▲	7.83E-02▲
$f(x)_{25}$	2.52E-11▲	2.52E-11▲	2.52E-11▲	2.52E-11▲	2.52E-11▲	2.52E-11▲	2.57E-10▲
$f(x)_{26}$	3.02E-11▲	3.02E-11▲	3.02E-11▲	3.02E-11▲	3.02E-11▲	3.02E-11▲	1.95E-03▲
$f(x)_{27}$	2.37E-12▲	2.37E-12▲	2.37E-12▲	2.37E-12▲	2.37E-12▲	2.37E-12▲	6.03E-11▲
$f(x)_{28}$	1.21E-12▲	1.21E-12▲	1.21E-12▲	1.21E-12▲	1.21E-12▲	1.21E-12▲	1.21E-12▲
▲	20	25	23	28	24	25	24
▼	4	3	4	0	3	3	4
►	4	0	1	0	1	0	0

compared to the MFO method, ES-EGT is superior in 24 functions and inferior in 3 and 1 functions, and the difference in performance is statisticaly inconclusive. In its comparison with the CSA method, ES-EGT is better on 25 functions and worse on 3. Finally, when compared to the OBSC method, ES-EGT shows better performance on 24 functions and is outperformed on 4. These results highlight, overall, the robustness and adaptability of the ES-EGT algorithm, especially when handling the complexities inherent in higher dimensional optimization tasks.

4.5.5 Comparison of Results on 100 Dimensions

In this section, we conduct a thorough evaluation of various metaheuristic algorithms, assessing their performance on a set of 28 functions extended to 100 dimensions. Handling functions in such a high-dimensional space poses a significantly more complex challenge in terms of finding solutions and optimizing them. The comprehensive test results are compiled in Table 4.5, which presents the performance outcomes in the context of 100 dimensions, analyzed through three key indices: Best Average (AB), Best Median (MD), and Standard Deviation (SD). From the data in Table 4.5, it is evident that the proposed ES-EGT algorithm excels in performance. Notably, the ES-EGT algorithm achieves superior solutions in 20 of the 28 functions tested. For 5 functions, its performance matches that of at least one other algorithm, while in only 3 functions, the ES-EGT algorithm is outperformed. These results showcase the effectiveness and adaptability of the ES-EGT method, particularly when dealing with the complexities of higher-dimensional optimization tasks. A notable observation from Table 4.5 is the ES-EGT algorithm's comparative performance improvement over other methods. This is especially evident in the increased number of functions where the ES-EGT algorithm outperforms its counterparts. This indicates the robustness of the ES-EGT algorithm, demonstrating its ability to maintain and even enhance its effectiveness in a highly challenging 100-dimensional environment. The consistency in outperforming other methods on most functions highlights the sophisticated design of the algorithm and its suitability for addressing complex, high-dimensional optimization problems.

The data presented in Table 4.5 reveal a clear hierarchy of performance among the algorithms tested in the 100-dimensional scenario, with the OBSC (Onlooker Bee Search for Modeling) algorithm ranking as the second best. This ranking is based on the results produced by the OBSC algorithm, which showed a commendable level of efficiency and effectiveness, distinguishing it from its competitors. The other algorithms in the study generally performed worse. This disparity in results is significant, as it underscores the varying capabilities of different metaheuristic approaches when faced with the increasing complexity of higher-dimensional search spaces. The ability of the OBCS algorithm to secure the second position highlights its robust algorithmic structure and its suitability for solving more intricate optimization problems. On the other hand, the less favorable results of the other algorithms indicate the possible limitations or challenges they face with the higher complexity of 100-dimensional environments. The comparison in Table 4.5 provides valuable information on the strengths and weaknesses of the different metaheuristic methods.

The algorithm's ability to efficiently tackle functions as complex as those defined in 100 dimensions is not solely due to its advanced fitting strategies but also the application of mechanisms that adeptly balance exploration and exploitation. A key example of such mechanisms is Eq. 4.9, specifically the exponential function e^{a}. Despite its apparent simplicity, this function plays a crucial role in dividing the search process into two distinct phases: the first 20% is dedicated to exploration, while the remaining 80% focuses on exploitation. During the initial 20% phase, the exponential

Table 4.5 Minimization results for 28 benchmark functions in 100 dimensions

Function		PSO	HS	BA	SMS	MFO	CSA	OBCS	ES-EGT
$f(x)_1$	AB	5.06E+00	1.83E+01	1.59E+01	2.13E+01	1.92E+01	1.88E+01	4.87E+00	**0.00E+00**
	MD	2.30E+00	1.83E+01	1.57E+01	2.13E+01	1.91E+01	1.88E+01	9.29E-01	**0.00E+00**
	SD	4.09E+00	1.68E-01	7.20E-01	5.08E-02	4.33E-01	1.52E-01	6.06E+00	**0.00E+00**
$f(x)_2$	AB	2.18E+05	4.26E+06	**9.12E-01**	5.72E+07	1.62E+06	5.23E+06	4.70E+05	9.90E-01
	MD	1.17E+05	4.35E+06	**7.68E-01**	5.64E+07	1.36E+06	5.31E+06	5.47E+05	1.00E+00
	SD	2.95E+05	4.16E+05	5.42E-01	3.01E+06	1.20E+06	5.60E+05	3.06E+05	**4.99E-02**
$f(x)_3$	AB	1.52E+01	7.93E+02	5.91E+02	3.63E+03	1.85E+02	9.69E+02	2.03E+01	**0.00E+00**
	MD	6.20E-02	7.96E+02	5.56E+02	3.64E+03	1.81E+02	9.71E+02	7.68E-01	**0.00E+00**
	SD	4.17E+01	3.43E+01	1.59E+02	7.72E+01	1.25E+02	3.73E+01	5.99E+01	**0.00E+00**
$f(x)_4$	AB	**0.00E+00**	1.93E+00	1.92E-14	4.18E+01	3.48E-01	2.00E+00	1.95E-02	**0.00E+00**
	MD	**0.00E+00**	2.01E+00	1.88E-14	4.14E+01	8.64E-05	2.10E+00	3.70E-09	**0.00E+00**
	SD	**0.00E+00**	2.88E-01	4.75E-15	2.90E+00	6.20E-01	4.69E-01	4.03E-02	**0.00E+00**
$f(x)_5$	AB	6.14E+01	2.58E+02	4.24E+01	1.74E+03	1.44E+02	3.61E+02	4.14E+01	**9.45E+00**
	MD	5.97E+01	2.59E+02	4.23E+01	1.73E+03	1.37E+02	3.60E+02	**1.36E+00**	9.48E+00
	SD	2.32E+01	1.68E+01	1.09E+01	8.59E+01	3.90E+01	1.98E+01	6.23E+01	**1.10E-01**
$f(x)_6$	AB	4.15E-06	5.94E-02	4.29E-14	3.16E+01	5.61E-05	3.22E-01	2.86E-02	**9.98E-294**
	MD	**0.00E+00**	6.01E-02	3.42E-14	3.13E+01	**0.00E+00**	3.26E-01	2.72E-02	**0.00E+00**
	SD	2.27E-05	8.67E-03	2.12E-14	1.75E+00	1.72E-04	3.74E-02	9.27E-03	**0.00E+00**
$f(x)_7$	AB	**0.00E+00**	3.78E-34	1.89E+25	1.62E+70	**0.00E+00**	2.09E+20	4.39E-08	**0.00E+00**
	MD	**0.00E+00**	1.11E-38	3.54E+11	4.01E+68	**0.00E+00**	6.28E+18	3.37E-25	**0.00E+00**
	SD	**0.00E+00**	1.41E-33	8.46E+25	4.66E+70	**0.00E+00**	5.84E+20	2.40E-07	**0.00E+00**
$f(x)_8$	AB	6.55E+04	6.55E+04	6.55E+04	6.55E+04	6.55E+04	6.55E+04	6.55E+04	**2.28E+04**

(continued)

Table 4.5 (continued)

Function		PSO	HS	BA	SMS	MFO	CSA	OBCS	ES-EGT
	MD	6.55E+04	6.55E+04	6.55E+04	6.55E+04	6.55E+04	6.55E+04	6.55E+04	**2.18E+04**
	SD	**0.00E+00**	**0.00E+00**	**0.00E+00**	**0.00E+00**	**0.00E+00**	**0.00E+00**	**0.00E+00**	5.90E+03
$f(x)_9$	AB	4.61E+01	1.22E+02	1.22E+02	3.25E+02	5.93E+01	1.60E+02	4.23E+01	**3.09E+01**
	MD	4.60E+01	1.23E+02	1.22E+02	3.25E+02	6.00E+01	1.61E+02	**3.00E+01**	**3.00E+01**
	SD	6.40E+00	3.27E+00	1.48E+01	7.05E+00	1.43E+01	4.18E+00	2.53E+01	**1.68E+00**
$f(x)_{10}$	AB	7.07E+01	3.46E+02	2.07E+02	3.81E+03	2.27E+02	4.06E+02	7.70E+01	**4.02E+01**
	MD	6.72E+01	3.51E+02	1.82E+02	3.78E+03	1.96E+02	4.10E+02	7.94E+01	**4.04E+01**
	SD	1.84E+01	2.93E+01	8.10E+01	1.50E+02	1.02E+02	3.86E+01	3.41E+01	**1.07E+00**
$f(x)_{11}$	AB	4.46E+02	8.27E+02	2.40E+02	2.08E+03	6.76E+02	1.12E+03	7.60E+02	**0.00E+00**
	MD	4.30E+02	8.28E+02	2.40E+02	2.09E+03	6.87E+02	1.13E+03	7.89E+02	**0.00E+00**
	SD	6.72E+01	1.94E+01	5.61E+01	4.75E+01	9.34E+01	2.51E+01	1.34E+02	**0.00E+00**
$f(x)_{12}$	AB	1.29E+06	1.09E+06	**1.47E+02**	1.82E+07	8.09E+05	2.33E+06	2.01E+04	1.10E+05
	MD	1.30E+06	1.09E+06	**1.43E+02**	1.81E+07	6.29E+05	2.31E+06	3.06E+02	1.16E+05
	SD	2.95E+05	1.14E+05	**5.67E+01**	8.58E+05	5.38E+05	2.90E+05	4.78E+04	3.75E+04
$f(x)_{13}$	AB	5.72E+01	7.58E+01	5.07E+01	1.00E+02	9.15E+01	7.07E+01	1.81E+01	**1.09E-173**
	MD	5.67E+01	7.60E+01	5.16E+01	1.00E+02	9.17E+01	7.09E+01	5.87E+00	**0.00E+00**
	SD	3.93E+00	1.16E+00	8.30E+00	0.00E+00	2.21E+00	2.32E+00	2.20E+01	**0.00E+00**
$f(x)_{14}$	AB	2.11E+03	5.34E+62	2.85E+118	1.01E+168	1.75E+03	2.00E+117	2.03E+89	**1.29E-178**
	MD	2.12E+03	2.50E+59	2.90E+107	3.38E+165	1.75E+03	6.80E+115	1.08E+67	**0.00E+00**
	SD	2.55E+02	1.58E+63	1.56E+119	6.55E+04	4.89E+02	6.22E+117	9.69E+89	**0.00E+00**
$f(x)_{15}$	AB	−2.19E+04	−2.32E+04	6.55E+04	4.79E+03	**−2.56E+04**	−1.54E+04	−1.84E+04	−5.08E+03
	MD	−2.21E+04	−2.32E+04	6.55E+04	4.36E+03	**−2.51E+04**	−1.55E+04	−1.83E+04	−4.99E+03

(continued)

Table 4.5 (continued)

Function		PSO	HS	BA	SMS	MFO	CSA	OBCS	ES-EGT
	SD	1.87E+03	5.73E+02	**0.00E+00**	1.03E+03	3.14E+03	1.21E+03	1.18E+03	6.35E+02
$f(x)_{16}$	AB	3.03E+03	8.58E+04	6.81E+04	4.06E+05	2.18E+04	1.05E+05	3.59E+03	**1.07E+00**
	MD	2.60E+01	8.66E+04	6.80E+04	4.05E+05	2.01E+04	1.07E+05	2.50E+00	**0.00E+00**
	SD	5.96E+03	5.92E+03	1.92E+04	1.43E+04	1.22E+04	5.96E+03	9.16E+03	**1.76E+00**
$f(x)_{17}$	AB	6.81E+07	9.30E+08	6.54E+08	4.49E+09	2.00E+08	9.22E+08	4.13E+06	**9.97E+01**
	MD	5.89E+07	9.26E+08	6.37E+08	4.47E+09	1.74E+08	9.32E+08	1.39E+06	**1.00E+02**
	SD	4.44E+07	6.00E+07	1.88E+08	2.16E+08	1.13E+08	6.29E+07	5.26E+06	**5.26E-01**
$f(x)_{18}$	AB	−7.78E+01	−6.91E+01	−6.11E+14	1.41E+01	**−9.17E+01**	−1.49E+14	−7.06E+01	−7.79E+00
	MD	−7.90E+01	−6.90E+01	−6.15E+14	1.47E+01	**−9.23E+01**	−1.51E+14	−6.96E+01	−7.03E+00
	SD	5.54E+00	**1.32E+00**	1.68E+14	2.84E+00	3.33E+00	3.04E+13	7.03E+00	4.07E+00
$f(x)_{19}$	AB	3.15E+03	1.37E+03	6.95E+03	**2.13E+01**	2.12E+03	1.46E+09	1.36E+03	6.22E+10
	MD	3.06E+03	1.38E+03	5.48E+03	**2.13E+01**	2.11E+03	9.53E+03	1.56E+03	5.02E+09
	SD	4.19E+02	9.44E+01	5.23E+03	**5.08E-02**	2.70E+02	7.95E+09	6.15E+02	1.80E+11
$f(x)_{20}$	AB	2.01E+05	1.54E+06	6.04E+05	9.02E+06	5.16E+05	2.14E+06	1.65E+04	**2.45E-267**
	MD	1.83E+05	1.54E+06	6.06E+05	9.04E+06	4.30E+05	2.14E+06	1.43E-01	**0.00E+00**
	SD	1.22E+05	8.49E+04	4.50E+05	2.64E+05	2.73E+05	1.41E+05	4.67E+04	**0.00E+00**
$f(x)_{21}$	AB	4.55E+07	1.89E+08	1.27E+08	1.47E+09	9.58E+07	2.90E+08	4.68E+07	**0.00E+00**
	MD	4.31E+07	1.92E+08	1.32E+08	1.46E+09	7.94E+07	2.94E+08	5.72E+07	**0.00E+00**
	SD	1.86E+07	1.39E+07	5.42E+07	6.27E+07	5.96E+07	2.04E+07	3.27E+07	**0.00E+00**
$f(x)_{22}$	AB	6.99E+00	2.23E+02	1.69E-04	1.08E+03	5.19E+01	2.79E+02	1.05E+01	**0.00E+00**
	MD	1.47E-04	2.21E+02	1.67E-04	1.08E+03	5.25E+01	2.80E+02	1.96E-06	**0.00E+00**
	SD	1.37E+01	1.41E+01	1.09E-05	2.96E+01	4.21E+01	1.29E+01	2.36E+01	**0.00E+00**

(continued)

Table 4.5 (continued)

Function		PSO	HS	BA	SMS	MFO	CSA	OBCS	ES-EGT
$f(x)_{23}$	AB	4.06E+03	3.61E+04	2.84E+00	2.12E+05	1.42E+04	5.01E+04	9.31E+03	**1.09E-296**
	MD	3.75E+03	3.65E+04	7.06E-02	2.11E+05	1.40E+04	5.01E+04	1.17E+04	**0.00E+00**
	SD	2.40E+03	1.77E+03	1.19E+01	8.52E+03	6.41E+03	2.72E+03	6.10E+03	**0.00E+00**
$f(x)_{24}$	AB	6.08E-12	8.27E-04	2.89E-09	6.08E+01	1.03E-07	1.07E-02	**0.00E+00**	**0.00E+00**
	MD	1.00E-13	8.04E-04	2.69E-09	6.05E+01	5.49E-11	9.85E-03	**0.00E+00**	**0.00E+00**
	SD	2.22E-11	2.19E-04	1.28E-09	2.48E+00	4.23E-07	5.30E-03	**0.00E+00**	**0.00E+00**
$f(x)_{25}$	AB	5.22E+04	8.30E+04	1.52E+22	2.50E+68	1.31E+05	1.61E+17	6.46E+03	**7.27E-240**
	MD	5.02E+04	8.48E+04	2.00E+07	2.69E+66	1.21E+05	3.80E+15	1.16E-01	**0.00E+00**
	SD	1.98E+04	6.52E+03	8.31E+22	8.64E+68	4.93E+04	4.38E+17	1.45E+04	**0.00E+00**
$f(x)_{26}$	AB	4.96E+02	3.20E+03	1.82E+03	1.29E+04	1.23E+03	3.60E+03	4.61E+02	**9.90E+01**
	MD	3.96E+02	3.20E+03	1.84E+03	1.29E+04	1.25E+03	3.62E+03	1.24E+02	**9.90E+01**
	SD	2.09E+02	1.03E+02	4.32E+02	4.86E+02	4.69E+02	1.75E+02	4.47E+02	**4.01E-05**
$f(x)_{27}$	AB	4.19E+03	7.70E+08	8.19E+07	1.01E+10	7.08E+07	8.61E+08	2.34E+06	**1.09E+02**
	MD	4.15E+03	7.81E+08	4.37E+07	1.01E+10	1.33E+05	8.58E+08	1.30E+04	**1.09E+02**
	SD	7.87E+02	8.76E+07	1.06E+08	4.11E+08	1.55E+08	8.62E+07	6.48E+06	**2.13E-06**
$f(x)_{28}$	AB	3.31E+03	3.23E+03	1.90E+19	1.73E+69	3.79E+03	4.21E+17	1.14E+03	**9.90E+01**
	MD	3.00E+03	3.23E+03	8.01E+07	2.54E+66	3.37E+03	2.58E+16	7.79E+02	**9.90E+01**
	SD	9.75E+02	1.22E+02	1.04E+20	7.17E+69	1.67E+03	1.28E+18	1.20E+03	**0.00E+00**

function generates relatively high values, facilitating significant jumps or movements in the solution space. This approach is essential for broadly exploring the vast and complex 100-dimensional landscape, enabling the algorithm to identify diverse and promising areas that might contain optimal or near-optimal solutions. In the final 80% phase, the focus shifts to exploitation. The exponential function yields smaller values during this phase, which is critical for refining and fine-tuning solutions. This phase is crucial for thoroughly analyzing the most promising areas identified during the exploration phase and progressively enhancing solutions to converge toward the optimal result. The effectiveness of this dual-phase approach lies in its ability to cover vast areas of the search space initially and then meticulously concentrate on the most promising regions. By balancing the broad scope of exploration with the detailed focus of exploitation, the algorithm exhibits a high degree of efficiency and effectiveness in navigating the complexities of high-dimensional optimization tasks.

4.5.6 Wilcoxon's Test Analysis on 100 Dimensions

Table 4.6 in the study introduces an important statistical perspective to the analysis, using p-values obtained from the Wilcoxon test. These p-values are calculated in relation to the performance data presented in Table 4.5, which focuses on the 100-dimensional case ($d = 100$). The essence of Table 4.6 lies in its facilitation of a comprehensive pairwise comparison of the ES-EGT algorithm with each of the other algorithms under test. This systematic approach of comparing the ES-EGT algorithm with others based on statistically significant metrics plays a key role in the objective validation of the experimental results.

The inclusion of these p-values in Table 4.6 is not a mere statistical exercise. However, it provides a solid basis for drawing reliable conclusions about the comparative effectiveness of the ES-EGT algorithm. Doing so reinforces the credibility of the experimental results and highlights the relative strengths and weaknesses of the various algorithms examined in a high-dimensional optimization context.

Table 4.6 presents a comprehensive statistical analysis using p-values from the Wilcoxon test, comparing the ES-EGT algorithm's performance with other methods in a 100-dimensional setting. Compared to the Particle Swarm Optimization (PSO) method, ES-EGT demonstrates statistically superior results in 23 functions, inferior results in 3 functions, and no definitive statistical winner in 2 functions. When compared with the Harmony Search (HS) method, ES-EGT outperforms 25 functions but lags in 3. Against the Bat Algorithm (BA), ES-EGT once again shows dominance in 23 functions while being less efficient in 5 functions. A noteworthy achievement is observed when comparing with the State of Matter Search (SMS) method, where ES-EGT statistically outperforms all 28 functions.

In comparison with the Moth-flame Optimization (MFO) method, ES-EGT excels in 24 functions, is outperformed in 3, and has statistically inconclusive results in 1 function. Compared to the Crow Search Algorithm (CSA), ES-EGT is superior in 25 functions and inferior in 3. Finally, compared to the Onlooker Bee Search for

Table 4.6 Results of Wilcoxon analysis for 28 test functions in 100 dimensions

	ES-EGT VS PSO	ES-EGT VS HS	ES-EGT VS BA	ES-EGT VS SMS	ES-EGT VS MFO	ES-EGT VS CSA	ES-EGT VS OBSC
$f(x)_1$	1.21E-12▲	1.21E-12▲	1.21E-12▲	1.21E-12▲	1.21E-12▲	1.21E-12▲	1.21E-12▲
$f(x)_2$	1.62E-11▲	1.62E-11▲	1.21E-04▼	1.62E-11▲	1.62E-11▲	1.62E-11▲	1.62E-11▲
$f(x)_3$	1.21E-12▲	1.21E-12▲	1.21E-12▲	1.21E-12▲	1.21E-12▲	1.21E-12▲	5.77E-11▲
$f(x)_4$	1.00E+00►	1.21E-12▲	1.21E-12▲	1.21E-12▲	1.21E-12▲	1.21E-12▲	4.57E-12▲
$f(x)_5$	3.02E-11▲	3.02E-11▲	3.02E-11▲	3.02E-11▲	3.02E-11▲	3.02E-11▲	1.91E-01▲
$f(x)_6$	1.00E+00▲	1.72E-12▲	1.72E-12▲	1.71E-12▲	3.99E-02▲	1.72E-12▲	1.72E-12▲
$f(x)_7$	1.00E+00►	1.21E-12▲	1.21E-12▲	1.21E-12▲	1.00E+00►	1.21E-12▲	6.25E-10▲
$f(x)_8$	1.21E-12▲	1.21E-12▲	1.21E-12▲	1.21E-12▲	1.21E-12▲	1.21E-12▲	1.21E-12▲
$f(x)_9$	2.02E-11▲	1.37E-11▲	1.41E-11▲	1.39E-11▲	2.49E-11▲	1.38E-11▲	4.40E-01▲
$f(x)_{10}$	3.02E-11▲	3.02E-11▲	3.02E-11▲	3.02E-11▲	3.02E-11▲	3.02E-11▲	3.83E-06▲
$f(x)_{11}$	1.21E-12▲	1.21E-12▲	1.21E-12▲	1.21E-12▲	1.21E-12▲	1.21E-12▲	1.21E-12▲
$f(x)_{12}$	3.02E-11▲	3.02E-11▲	3.02E-11▼	3.02E-11▲	3.02E-11▲	3.02E-11▲	1.36E-07▼
$f(x)_{13}$	9.40E-12▲	9.40E-12▲	9.40E-12▲	2.62E-13▲	9.40E-12▲	9.40E-12▲	9.40E-12▲
$f(x)_{14}$	1.96E-11▲	1.96E-11▲	1.96E-11▲	1.96E-11▲	1.96E-11▲	1.96E-11▲	1.96E-11▲
$f(x)_{15}$	3.02E-11▼	3.02E-11▼	1.21E-12▼	3.02E-11▲	3.02E-11▼	3.02E-11▼	3.02E-11▼
$f(x)_{16}$	1.43E-11▲	1.43E-11▲	1.43E-11▲	1.43E-11▲	1.43E-11▲	1.43E-11▲	2.21E-02▲
$f(x)_{17}$	3.02E-11▲	3.02E-11▲	3.02E-11▲	3.02E-11▲	3.02E-11▲	3.02E-11▲	5.57E-10▲
$f(x)_{18}$	3.02E-11▼	3.02E-11▼	9.65E-06▼	3.02E-11▲	3.02E-11▼	1.85E-10▼	3.02E-11▼
$f(x)_{19}$	3.02E-11▼	3.02E-11▼	3.02E-11▼	3.02E-11▲	3.02E-11▼	5.57E-10▼	3.02E-11▼
$f(x)_{20}$	2.37E-12▲	2.37E-12▲	2.37E-12▲	2.37E-12▲	2.37E-12▲	2.37E-12▲	2.37E-12▲
$f(x)_{21}$	1.72E-12▲	1.72E-12▲	1.72E-12▲	1.72E-12▲	1.72E-12▲	1.72E-12▲	1.72E-12▲
$f(x)_{22}$	1.21E-12▲	1.21E-12▲	1.21E-12▲	1.21E-12▲	1.21E-12▲	1.21E-12▲	4.57E-12▲
$f(x)_{23}$	1.72E-12▲	1.72E-12▲	1.72E-12▲	1.72E-12▲	1.72E-12▲	1.72E-12▲	1.72E-12▲
$f(x)_{24}$	1.21E-12▲	1.21E-12▲	1.21E-12▲	1.11E-12▲	1.21E-12▲	1.21E-12▲	1.00E+00►
$f(x)_{25}$	1.10E-11▲	1.10E-11▲	1.10E-11▲	1.10E-11▲	1.10E-11▲	1.10E-11▲	1.10E-11▲
$f(x)_{26}$	3.02E-11▲	3.02E-11▲	3.02E-11▲	3.02E-11▲	3.02E-11▲	3.02E-11▲	7.96E-03▲
$f(x)_{27}$	2.92E-11▲	2.92E-11▲	2.92E-11▲	2.92E-11▲	2.92E-11▲	2.92E-11▲	2.92E-11▲
$f(x)_{28}$	1.21E-12▲	1.21E-12▲	1.21E-12▲	1.21E-12▲	1.21E-12▲	1.21E-12▲	1.21E-12▲
▲	23	25	23	28	24	25	23
▼	3	3	5	0	3	3	4
►	2	0	0	0	1	0	1

Modeling (OBSC) method, ES-EGT shows better performance in 23 functions and superior results in 4 functions, with one function showing statistically indeterminate results. These findings highlight the robustness and adaptability of the ES-EGT algorithm, particularly in handling the complexities of high-dimensional optimization tasks. The algorithm's consistently high performance across most functions underscores its effectiveness and sophistication in addressing complex, high-dimensional problems.

4.5.7 Convergence Analysis

Convergence analysis in metaheuristic methods involves examining how an algorithm advances and reaches an optimal or near-optimal solution through iterations. The primary goal of this analysis is to assess a metaheuristic algorithm's efficiency and effectiveness, especially in achieving a high-quality solution within a practical number of iterations. In this subsection, we concentrate on comparing the convergence behavior of the proposed ES-EGT method with seven established metaheuristic algorithms, including PSO, HS, BA, SMS, MFO, CSA, and OBSC. This comparative analysis aims to showcase the relative strengths of ES-EGT in achieving convergence and its performance against other prominent algorithms in the field.

The convergence analysis performed in this study utilized two representative functions for each of the three-dimensional scenarios: 30, 50, and 100 dimensions, to provide a focused and relevant assessment of algorithmic performance. For the 30-dimensional case, functions $f(x)_2$ and $f(x)_{10}$ were selected due to their distinctive characteristics and relevance to this dimensionality. For the 50-dimensional scenario, functions $f(x)_{11}$ and $f(x)_{13}$ were chosen, each presenting unique challenges and complexities suitable for evaluating algorithms in a moderately high-dimensional space. Finally, for the more complex 100-dimensional scenario, functions $f(x)_{19}$ and $f(x)_{21}$ were selected for their ability to test the limits of algorithms in handling high-dimensional optimization tasks. By using these specific functions across different dimensions, the analysis aims to provide a comprehensive understanding of each algorithm's performance in terms of convergence under varying levels of problem complexity.

Figure 4.6 of the study presents the convergence plots of all the algorithms included in the comparative analysis. From this visualization, it is evident that the proposed ES-EGT algorithm excels, achieving the best solution faster than its counterparts. This rapid convergence to the optimal solution underscores the efficiency and effectiveness of the ES-EGT algorithm in navigating the search space. The OBSC (Onlooker Bee Search for Modeling) algorithm follows closely, securing the second position in terms of the speed at which it finds the best solution. This outstanding performance highlights its robustness in convergence among the evaluated algorithms.

In contrast, the other algorithms in the comparison exhibit similar performance levels. This relative uniformity among the other algorithms emphasizes the capability of the ES-EGT and OBSC algorithms to converge quickly and efficiently to the best solutions within their respective search spaces. Thus, Fig. 4.6 provides a clear comparative perspective of each algorithm's convergence capabilities, with the ES-EGT algorithm demonstrating a significant advantage in achieving rapid and efficient convergence.

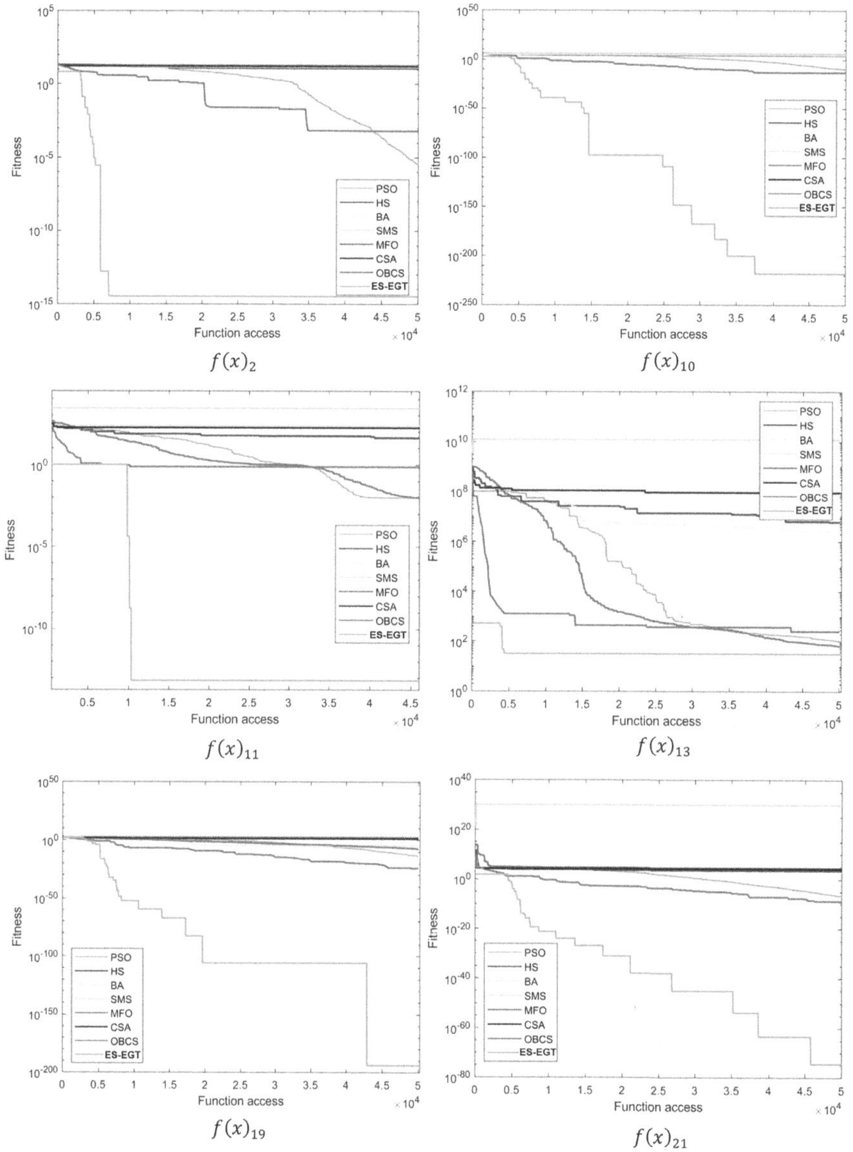

Fig. 4.6 Convergence plots of all the algorithms included in the comparative analysis

The exceptional effectiveness of the proposed algorithm in attaining fast convergence times is attributed to its innovative modification of the Evolutionary Game Theory (EGT) adaptation mechanism. In the standard EGT framework, strategy modifications are performed through pairwise comparisons among agents, where the worst-performing agent adapts its strategy to resemble that of the best-performing agent more closely. However, this traditional approach typically requires numerous iterations for the optimal strategy to emerge and dominate the entire population of agents. Our modified approach enhances the EGT mechanism by introducing a third element into these pairwise comparisons. This third element represents the agent with the best overall strategy, identified based on the most satisfactory search results. By incorporating this top-performing agent into the strategy adaptation process, our algorithm accelerates convergence to the optimal strategy. This innovative modification allows the algorithm to quickly identify and adopt the most efficient strategies in far fewer iterations than the original EGT mechanism would require. Consequently, this approach significantly reduces the time needed for the entire population of agents to converge on the most efficient and effective strategy, thereby improving the overall speed and efficiency of the optimization process.

4.6 Conclusions

This chapter introduces a new metaheuristic algorithm that dynamically fine-tunes its search strategy throughout the optimization process to enhance its efficiency. The foundation of this algorithm lies in Evolutionary Strategies (ES), selected for their exceptional self-adaptive properties. To further boost its effectiveness, the algorithm integrates aspects of Evolutionary Game Theory (EGT). This integration broadens the strategy adaptation process by not only considering information from the specific agent but also incorporating insights from other members of the population. Our approach also innovatively modifies the traditional EGT mechanism. We extend beyond standard peer evaluations and incorporate data from the top-performing individuals within the population, factoring in their scores. This thorough strategy adaptation enables the algorithm to converge on the most effective and dominant strategy quickly.

To assess the effectiveness of our new method, we conducted an extensive analysis, comparing it with various established metaheuristic algorithms such as PSO, HS, BA, SMS, MFO, CSA, and OBSC. In this evaluation, we utilized 28 diverse test functions chosen to represent a wide range of optimization challenges. The results were highly enlightening, demonstrating that our methodology significantly outperforms existing techniques. It not only delivers higher quality solutions but does so at a faster rate. This rapid convergence, combined with the superior quality of the solutions, underscores the effectiveness of our approach and signifies a major advancement in metaheuristic algorithms. This analysis confirms our method's ability to efficiently solve complex optimization problems, offering a robust and viable alternative to conventional algorithms.

Appendix: List of Benchmark Functions

	Name	Global minimum	Boundaries	Function
$f(x)_1$	Ackley	$f(\mathbf{x}^*) = 0$; $\mathbf{x}^* = (0, \ldots, 0)$	$[-30, 30]^d$	$f(x) = -20\exp\left(-0.2\sqrt{\frac{1}{d}\sum_{i=1}^{d} x_i^2}\right) - exp\left(\frac{1}{d}\sum_{i=1}^{d}\cos(2\pi x_i)\right) + 20 + \exp(1)$
$f(x)_2$	Dixon-Price	$f(\mathbf{x}^*) = 0$; $\mathbf{x}^* = 2^{-\frac{2^i-2}{2^i}}$ for $i = 1, \ldots, n$	$[-10, 10]^d$	$f(x) = (x_1 - 1)^2 + \sum_{i=1}^{d} i\left(2x_i^2 - x_{i-1}\right)^2$
$f(x)_3$	Griewank	$f(\mathbf{x}^*) = 0$; $\mathbf{x}^* = (0, \ldots, 0)$	$[-600, 600]^d$	$f(x) = \sum_{i=1}^{d} \frac{x_i^2}{4000} - \prod_{i=1}^{d} \cos\left(\frac{x_i}{\sqrt{i}}\right) + 1$
$f(x)_4$	Infinity	$f(\mathbf{x}^*) = 0$; $\mathbf{x}^* = (0, \ldots, 0)$	$[-1, 1]^d$	$f(x) = \sum_{i=1}^{d} x_i^6 sen(x_i + 2)$
$f(x)_5$	Levy	$f(\mathbf{x}^*) = 0$; $\mathbf{x}^* = (1, \ldots, 1)$	$[-10, 10]^d$	$f(x) = \sin^2(\pi\omega_1) + \sum_{i=1}^{d-1}(\omega_1 - 1)^2\left[1 + 10\sin^2(\pi\omega_i + 1)\right] + (\omega_d - 1)^2\left[1 + \sin^2(2\pi\omega_d)\right]$; $\omega_i = 1 + \frac{x_i - 1}{4}$
$f(x)_6$	Mishra 11	$f(\mathbf{x}^*) = 0$; $\mathbf{x}^* = (0, \ldots, 0)$	$[-10, 10]^d$	$f(x) = \left[\frac{1}{d}\sum_{i=1}^{d}\lvert x_i\rvert - \left(\prod_{i=1}^{d}\lvert x_i\rvert\right)^{\frac{1}{d}}\right]^2$
$f(x)_7$	Multimodal	$f(\mathbf{x}^*) = 0$; $\mathbf{x}^* = (-1, \ldots, -1)$	$[-10, 10]^d$	$f(x) = \sum_{i=1}^{d}\lvert x_i\rvert \prod_{i=1}^{d}\lvert x_i\rvert$
$f(x)_8$	Perm 2	$f(\mathbf{x}^*) = 0$; $\mathbf{x}^* = (1, 1/2, \ldots, 1/n)$	$[-d, d]^d$	$f(x) = \sum_{i=1}^{d}\left[\sum_{j=1}^{d}(j^i + 10)\left(x_j^i - \frac{1}{j^i}\right)\right]^2$
$f(x)_9$	Plateau	$f(\mathbf{x}^*) = 30$; $\mathbf{x}^* = (0, \ldots, 0)$	$[-5.12, 5.12]^d$	$f(x) = 30 + \sum_{i=1}^{d}\lvert x_i\rvert$
$f(x)_{10}$	Quartic	$f(\mathbf{x}^*) = 0$; $\mathbf{x}^* = (-1, \ldots, -1)$	$[-10, 10]^d$	$f(x) = \sum_{i=1}^{d} ix_i^4 + rand[0, 1]$
$f(x)_{11}$	Rastringin	$f(\mathbf{x}^*) = 0$; $\mathbf{x}^* = (1, \ldots, 1)$	$[-5, 10]^d$	$f(x) = 10d + \sum_{i=1}^{d}\left[x_i^2 - 10cos(2\pi x_i)\right]$
$f(x)_{12}$	Rosenbrock	$f(\mathbf{x}^*) = 0$; $\mathbf{x}^* = (0.5, \ldots, 0.5)$	$[-100, 100]^d$	$f(x) = \sum_{i=1}^{d} 100\left(x_{i+1} - x_i^2\right)^2 + (x_i - 1)^2$
$f(x)_{13}$	Schweffel 21	$f(\mathbf{x}^*) = 0$; $\mathbf{x}^* = (0, \ldots, 0)$	$[-100, 100]^d$	$f(x) = max\{\lvert x_i\rvert, 1 \le i \le d\}$
$f(x)_{14}$	Schweffel 22	$f(\mathbf{x}^*) = 0$; $\mathbf{x}^* = (0, \ldots, 0)$	$[-100, 100]^d$	$f(x) = \sum_{i=1}^{d}\lvert x_i\rvert + \prod_{i=1}^{d}\lvert x_i\rvert$
$f(x)_{15}$	Schweffel 26	$f(\mathbf{x}^*) = 0$; $\mathbf{x}^* = \begin{pmatrix} 420.9867, \ldots, \\ 420.9867 \end{pmatrix}$	$[-500, 500]^d$	$f(x) = \sum_{i=1}^{d}\left(-x_i\left(sin\left(\sqrt{\lvert x_i\rvert}\right)\right)\right)$
$f(x)_{16}$	Step	$f(\mathbf{x}^*) = 0$; $\mathbf{x}^* = (0, \ldots, 0)$	$[-100, 100]^d$	$f(x) = \sum_{i=1}^{d}\lvert x_i^2\rvert$
$f(x)_{17}$	Trid	$f(\mathbf{x}^*) = -\frac{n(n+4)(n-1)}{6}$; $\mathbf{x}^* = (i(n+1) - i)$; $i = 1, \ldots, n$	$[-d^2, d^2]^d$	$f(x) = \sum_{i=1}^{d}(x_i - 1)^2 - \sum_{i=2}^{d} x_i x_{i-1}$

(continued)

(continued)

	Name	Global minimum	Boundaries	Function
$f(x)_{18}$	Vincent	$f(\mathbf{x}^*) = -n$; $\mathbf{x}^* = (7.70, \ldots, 7.70)$;	$[0.25, 10]^d$	$f(x) = -\frac{1}{d}\sum_{i=1}^{d} sin[10\log(x_i)]$
$f(x)_{19}$	Zakharov	$f(\mathbf{x}^*) = 0$; $\mathbf{x}^* = (0, \ldots, 0)$;	$[-5, 10]^d$	$f(x) = \sum_{i=1}^{d} x_i^2 + \left(\sum_{i=1}^{d} 0.5ix_i\right)^2 + \left(\sum_{i=1}^{d} 0.5ix_i\right)^4$
$f(x)_{20}$	Rothyp	$f(\mathbf{x}^*) = 0$; $\mathbf{x}^* = (0, \ldots, 0)$;	$[-65.536, 65.536]^d$	$f(x) = \sum_{i=1}^{d}\sum_{j=1}^{i} x_j^2$
$f(x)_{21}$	Schweffel 2	$f(\mathbf{x}^*) = 0$; $\mathbf{x}^* = (0, \ldots, 0)$;	$[-100, 100]^d$	$f(x) = \sum_{i=1}^{d}\left(\sum_{j=1}^{i} x_i\right)^2$
$f(x)_{22}$	Sphere	$f(\mathbf{x}^*) = 0$; $\mathbf{x}^* = (0, \ldots, 0)$;	$[-100, 100]^d$	$f(x) = \sum_{i=1}^{d} x_i^2$
$f(x)_{23}$	Sum squares	$f(\mathbf{x}^*) = 0$; $\mathbf{x}^* = (0, \ldots, 0)$;	$[-10, 10]^d$	$f(x) = \sum_{i=1}^{d} ix_i^2$
$f(x)_{24}$	Sum powers	$f(\mathbf{x}^*) = 0$; $\mathbf{x}^* = (0, \ldots, 0)$;	$[-1, 1]^d$	$f(x) = \sum_{i=1}^{d} \lvert x_i\rvert^{i+1}$
$f(x)_{25}$	Rastringin + Schweffel 22 + Sphere	$f(\mathbf{x}^*) = 0$; $\mathbf{x}^* = (0, \ldots, 0)$;	$[-100, 100]^d$	$f(x) = \left[10d + \sum_{i=1}^{d}\left[x_i^2 - 10cos(2\pi x_i)\right]\right] + \left[\sum_{i=!}^{d}\lvert x_i\rvert + \prod_{i=1}^{d}\lvert x_i\rvert\right] + \left[\sum_{i=1}^{d} x_i^2\right]$
$f(x)_{26}$	Griewank + Rastringin + Rosenbrock	$f(\mathbf{x}^*) = n - 1$; $\mathbf{x}^* = (0, \ldots, 0)$;	$[-100, 100]^d$	$f(x) = \left[\sum_{i=1}^{d}\frac{x_i^2}{4000} - \prod_{i=1}^{d}\cos\left(\frac{x_i}{\sqrt{i}}\right) + 1\right] + \left[10d + \sum_{i=1}^{d}\left[x_i^2 - 10cos(2\pi x_i)\right]\right] + \left[\sum_{i=1}^{d} 100\left(x_{i+1} - x_i^2\right)^2 + (x_i - 1)^2\right]$
$f(x)_{27}$	Ackley + Penalty 2 + Rosenbrock + Schweffel 22	$f(\mathbf{x}^*) = (1.1n) - 1$; $\mathbf{x}^* = (0, \ldots, 0)$;	$[-100, 100]^d$	$f(x) = \left[-20\exp\left(-0.2\sqrt{\frac{1}{d}\sum_{i=1}^{d} x_i^2}\right) - \exp\left(\frac{1}{d}\sum_{i=1}^{d}\cos(2\pi x_i)\right) + 20 + \exp(1)\right] + \left[0.1 \times \{\sin^2(3\pi x_1) + \sum_{i=1}^{d-1}(x_i - 1)^2 \left[1 + \sin^2(3\pi x_{i+1})\right] + (x_d - 1)^2 [1 + \sin^2(2\pi x_d)]\} + \sum_{i=1}^{n} u(x_i, a, k, m)\right] + \left[\sum_{i=1}^{d} 100\left(x_{i+1} - x_i^2\right)^2 + (x_i - 1)^2\right] + \left[\sum_{i=!}^{d}\lvert x_i\rvert + \prod_{i=1}^{d}\lvert x_i\rvert\right]$

(continued)

(continued)

	Name	Global minimum	Boundaries	Function
$f(x)_{28}$	Ackley + Griewank + Rastringin + Rosenbrock + Schweffel 22	$f(\mathbf{x}^*) = n - 1$; $\mathbf{x}^* = (0, \ldots, 0)$;	$[-100, 100]^d$	$f(x) = \left[-20\exp\left(-0.2\sqrt{\frac{1}{d}\sum_{i=1}^{d} x_i^2}\right) - \exp\left(\frac{1}{d}\sum_{i=1}^{d}\cos(2\pi x_i)\right) + 20 + \exp(1)\right] + \left[\sum_{i=1}^{d}\frac{x_i^2}{4000} - \prod_{i=1}^{d}\cos\left(\frac{x_i}{\sqrt{i}}\right) + 1\right] + \left[10d + \sum_{i=1}^{d}\left[x_i^2 - 10cos(2\pi x_i)\right]\right] + \left[\sum_{i=1}^{d} 100\left(x_{i+1} - x_i^2\right)^2 + (x_i - 1)^2\right] + \left[\sum_{i=!}^{d}\lvert x_i\rvert + \prod_{i=1}^{d}\lvert x_i\rvert\right]$

References

1. P. Hammerstein, R. Selten, Game theory and evolutionary biology (Chap. 28) (1994), pp. 929–993. https://doi.org/10.1016/S1574-0005(05)80060-8
2. J.M. Smith, Game theory and the evolution of behaviour. Behav. Brain Sci. **7**(1), 95–101 (1984). https://doi.org/10.1017/S0140525X00026327
3. J.M. Smith, G.R. Price, The logic of animal conflict. Nature **246**(5427), 15–18 (1973). https://doi.org/10.1038/246015a0
4. E. Rasmusen, *Games and Information: An Introduction to Game Theory* (Blackwell Publishing, 2007)
5. R. Gibbons, *A Primer in Game Theory* (Pearson Education, 1992)
6. K.G. Binmore, *Game Theory and the Social Contract: Just Playing*, vol. 2 (MIT Press, 1994)
7. M. Shubik, *Game Theory, Law, and the Concept of Competition*, 285th edn., vol. 60 (1991)
8. O. Chatain, Cooperative and non-cooperative game theory, HEC Paris
9. D.M. Kreps, *Game Theory and Economic Modelling* (Oxford University Press, 1990)
10. A.J. Stewart, J.B. Plotkin, Extortion and cooperation in the Prisoner's Dilemma. Proc. Natl. Acad. Sci. **109**(26), 10134–10135 (2012). https://doi.org/10.1073/pnas.1208087109
11. B. Grofman, J. Pool, How to make cooperation the optimizing strategy in a two-person game. J. Math. Sociol. **5**(2), 173–186 (1977). https://doi.org/10.1080/0022250X.1977.9989871
12. C. Wedekind, M. Milinski, Human cooperation in the simultaneous and the alternating Prisoner's Dilemma: Pavlov versus generous tit-for-tat. Proc. Natl. Acad. Sci. **93**(7), 2686–2689 (1996). https://doi.org/10.1073/pnas.93.7.2686
13. J.M. Smith, *Evolution and the Theory of Games* (Cambridge University Press, 1982). https://doi.org/10.1017/CBO9780511806292
14. J. Maynard Smith, *Evolutionary Game Theory* (1980), pp. 73–81. https://doi.org/10.1007/978-3-642-93161-1_5

15. J.G. Riley, Evolutionary equilibrium strategies. J. Theor. Biol. **76**(2), 109–123 (1979). https://doi.org/10.1016/0022-5193(79)90365-5
16. J.M. Alexander, *Evolutionary Game Theory* (Cambridge University Press, 2023). https://doi.org/10.1017/9781108582063
17. S. Maynard, *Evolutionary Genetics*, 2nd edn. (Oxford University Press, 1998)
18. J.M. Smith, G.A. Parker, The logic of asymmetric contests. Anim. Behav. **24**(1), 159–175 (1976). https://doi.org/10.1016/S0003-3472(76)80110-8
19. E. Cuevas, A. Luque, N. Aguirre, M.A. Navarro, A. Rodríguez, Metaheuristic optimization with dynamic strategy adaptation: an evolutionary game theory approach. Phys. A **645**, 129831 (2024). https://doi.org/10.1016/j.physa.2024.129831
20. T. Back, *Evolutionary Algorithms in Theory and Practice: Evolution Strategies, Evolutionary Programming, Genetic Algorithms* (Oxford University Press, 1996)
21. K. Karabulut, H. Öztop, D. Kizilay, M.F. Tasgetiren, L. Kandiller, An evolution strategy approach for the distributed permutation flowshop scheduling problem with sequence-dependent setup times. Comput. Oper. Res. **142**, 105733 (2022). https://doi.org/10.1016/j.cor.2022.105733
22. C. Huang, X. Zhou, X. Ran, Y. Liu, W. Deng, W. Deng, Co-evolutionary competitive swarm optimizer with three-phase for large-scale complex optimization problem. Inf. Sci. (N. Y.) **619**, 2–18 (2023). https://doi.org/10.1016/j.ins.2022.11.019
23. G. Wang, Y. Chao, T. Jiang, Z. Chen, Facilitating developments of solar thermal power and nuclear power generations for carbon neutral: a study based on evolutionary game theoretic method. Sci. Total. Environ. **814**, 151927 (2022). https://doi.org/10.1016/j.scitotenv.2021.151927
24. J.-W. Lee, M.-K. Kim, An evolutionary game theory-based optimal scheduling strategy for multiagent distribution network operation considering voltage management. IEEE Access **10**, 50227–50241 (2022). https://doi.org/10.1109/ACCESS.2022.3174077
25. A. Traulsen, N.E. Glynatsi, The future of theoretical evolutionary game theory. Philos. Trans. R. Soc. B: Biol. Sci. **378**(1876) (2023). https://doi.org/10.1098/rstb.2021.0508
26. D. Challet, Y.-C. Zhang, Emergence of cooperation and organization in an evolutionary game. Phys. A **246**(3–4), 407–418 (1997). https://doi.org/10.1016/S0378-4371(97)00419-6
27. M. Scatà et al., Combining evolutionary game theory and network theory to analyze human cooperation patterns. Chaos Solitons Fractals **91**, 17–24 (2016). https://doi.org/10.1016/j.chaos.2016.04.018
28. B. Morales-Castañeda, D. Zaldívar, E. Cuevas, F. Fausto, A. Rodríguez, A better balance in metaheuristic algorithms: Does it exist? Swarm Evol. Comput. **54**, 100671 (2020). https://doi.org/10.1016/j.swevo.2020.100671
29. V. Osuna-Enciso, E. Cuevas, B. Morales Castañeda, A diversity metric for population-based metaheuristic algorithms. Inf. Sci. (N. Y.) **586**, 192–208 (2022). https://doi.org/10.1016/j.ins.2021.11.073
30. B. Morales-Castañeda, D. Zaldívar, E. Cuevas, A. Rodríguez, M.A. Navarro, Population management in metaheuristic algorithms: could less be more? Appl. Soft Comput. **107**, 107389 (2021). https://doi.org/10.1016/j.asoc.2021.107389
31. J. Kennedy, R. Eberhart, Particle swarm optimization, in *Proceedings of ICNN'95 - International Conference on Neural Networks* (IEEE), pp. 1942–1948. https://doi.org/10.1109/ICNN.1995.488968
32. Z.W. Geem, J.H. Kim, G.V. Loganathan, A new heuristic optimization algorithm: harmony search. Simulation **76**(2), 60–68 (2001). https://doi.org/10.1177/003754970107600201
33. X.-S. Yang, A new metaheuristic bat-inspired algorithm (2010), pp. 65–74. https://doi.org/10.1007/978-3-642-12538-6_6
34. X.S. Yang, X. He, Bat algorithm: literature review and applications. Int. J. Bio-Inspired Comput. **5**(3), 141 (2013). https://doi.org/10.1504/IJBIC.2013.055093
35. E. Cuevas, A. Echavarría, M.A. Ramírez-Ortegón, An optimization algorithm inspired by the states of matter that improves the balance between exploration and exploitation. Appl. Intell. **40**(2), 256–272 (2014). https://doi.org/10.1007/s10489-013-0458-0

36. S. Mirjalili, Moth-flame optimization algorithm: a novel nature-inspired heuristic paradigm. Knowl Based Syst **89**, 228–249 (2015). https://doi.org/10.1016/j.knosys.2015.07.006
37. A.G. Hussien et al., Crow search algorithm: theory, recent advances, and applications. IEEE Access **8**, 173548–173565 (2020). https://doi.org/10.1109/ACCESS.2020.3024108
38. X. Zhu, N. Wang, Cuckoo search algorithm with onlooker bee search for modeling PEMFCs using T2FNN. Eng. Appl. Artif. Intell. **85**, 740–753 (2019). https://doi.org/10.1016/j.engappai.2019.07.019
39. F. Wilcoxon, Probability tables for individual comparisons by ranking methods. Biometrics **3**(3), 119 (1947). https://doi.org/10.2307/3001946

Chapter 5
Learning Strategies in Metaheuristics

5.1 Introduction

Learning is an essential process in life; it encompasses the acquisition of knowledge through different personal factors and social factors [1]. Each of these factors plays an important role in collecting information and finding solutions to different challenges that may arise during the optimization process.

The ability to learn is not an act unique to human beings, as this peculiarity has also been seen in nature itself; for example, dogs can be trained to perform a specific task where they make use of their skills. Therefore, it has also been observed that the behavior of predators and prey has changed over the years because predators can learn about the behavior of others and thus modify their own to have more chances of being victorious. As the years went by, human beings acquired more and more knowledge and learned new skills, such as the ability to create increasingly sophisticated and complex inventions, such as artificial intelligence, which also has obtained the ability to learn; that is, learning is no longer unique to living beings, but machines and programs can also possess this learning ability. Consequently, it is possible for a programming algorithm to learn techniques that modify its behavior and thereby seek new possibilities in the search for results [2].

As a result, these algorithms can learn in the same ways as humans, under individual and collective factors. Each of these is known as a strategy [3] that seeks to help the program acquire knowledge and, thus, learn. However, each one focuses on different procedures to carry out this process successfully. One focuses on what the individual does and achieves individually. The other focuses on the interaction that this individual has with others and the knowledge that they possess.

E. Cuevas et al., *Advanced Metaheuristics: Novel Approaches for Complex Problem Solving*, Studies in Computational Intelligence 1210,
https://doi.org/10.1007/978-3-031-89284-4_5

5.2 Individual Learning Strategies (I-LS)

Individual Learning [4], also known as personal learning, is a learning process that is characterized by the direct interaction that the individual has with his surrounding environment. This type of interaction represents, very accurately, that learning can be obtained in various ways, as shown in Fig. 5.1, where it can be observed that knowledge can be acquired through experience, by performing a test and learning from mistakes and successes, sensory experiences (those where one or more senses participate) and direct participation, among others.

In metaheuristic algorithms, more specifically in particles moving through the search space, this learning can be observed as the process of exploration within the search space. This learning is reflected through specific tasks that must be completed, such as exploring a specific region within the space, where the particle learns from its direct results, by which it adopts variations within its work strategy with the purpose of improving and obtaining better results in the future. In Table 5.1, we can observe the advantages and disadvantages of this learning strategy in metaheuristic algorithms.

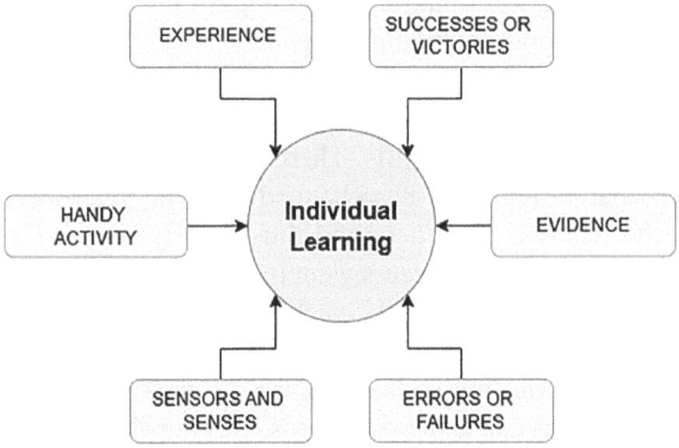

Fig. 5.1 Ways of learning individually

Table 5.1 Advantages and disadvantages of the I-LS in metaheuristics

ADVANTAGES	DISADVANTAGES
• A good exploitation of the answers	• Possibility of early stagnation
• Low computational cost and faster problem resolution	• Little exploration of results in the search space
• Positive results in unimodal problems	• Ineffective results in engineering problems

5.3 Collective Learning Strategies (C-LS)

Collective learning [5], also known as social learning, is a learning process that involves the interaction of an individual with others within the same environment. This process is a reference to how one can learn by observing and imitating others. Similar to individual learning and as shown in Fig. 5.2, collective learning can be achieved through the experience of other individuals, through criticism and feedback from other individuals with knowledge of the subject, or simply through the communication that exists between an individual and the rest.

In metaheuristic algorithms, more specifically search particles, this learning means the possibility of learning from the experience of other particles, from the information they may possess or from the strategies they use to achieve the desired optimal result. All of this can be observed in the process of imitation between these individuals, where a particle observes and repeats the actions of those it considers victorious or those that have a better performance compared to its own. In Table 5.2, we can observe the advantages and disadvantages of applying a Collective Learning Strategy within the metaheuristic algorithms. Thus, we can obtain an idea of the behavior that would be expected.

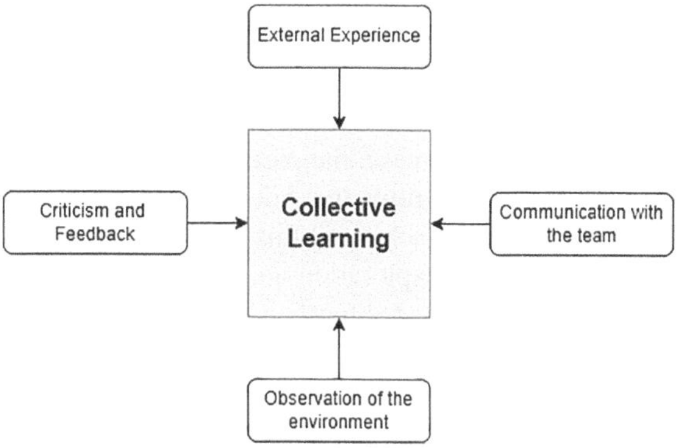

Fig. 5.2 Ways of learning collectively

Table 5.2 Advantages and disadvantages of the C-LS in metaheuristics

ADVANTAGES	DISADVANTAGES
• A good exploration of the search space	• Little exploitation of the results obtained
• Less chance of stagnation due to its continuous movement	• Slower problem-solving speed and thus a higher computational cost
• Positive results in multimodal problems	• Ineffective results in unimodal problems

5.4 Balance Between Individual and Collective Learning Strategies

Learning is a process that involves different factors such as those mentioned above, with which we can conclude that to obtain better learning, it is necessary to achieve a balance between these same factors, where the individual factor causes the individual to develop solutions according to his own experience in the environment. The collective factor causes the individual to have a better observation of the environment that surrounds him through the experiences of the rest of the individuals that surround him, where he obtains strategies depending on the victories and defeats of the rest of the population [6].

In the field of metaheuristic algorithms, the balance of these strategies refers to the frequency with which the individual will act according to his knowledge or act according to the information obtained from other individuals in the search space. This balance is essential because each of the strategies has its respective advantages and disadvantages, as observed in Tables 5.1 and 5.2; by combining the advantages, the disadvantages are eliminated so that the individual can obtain better solutions, and thus, the algorithm will perform better, obtaining more optimal solutions.

Without the existence of this balance, various consequences may be observed, such as the disadvantages mentioned, since the particle focuses more on its strategy regardless of whether it gives optimal solutions or not, causing it not to observe the rest of the population. Therefore, it will not have a better understanding of the search space, affecting the exploration process. Similarly, it focuses solely on the collective strategy. In that case, the particle will simply be imitating what the rest of the population does, resulting in it not having its own strategy, affecting the exploitation stage. Therefore, by achieving this equilibrium or balance, greater effectiveness may be obtained in the exploration and exploitation stages of the algorithm, which are considered crucial stages for the resolution of optimization problems.

5.5 Design and Implementation of the Algorithm

Metaheuristic algorithms have characteristic stages called operators [7], where said operators change depending on their classifications; for example, bio-inspired algorithms seek to behave like nature or animal instinct from which they obtain their inspiration, or evolutionary algorithms that seek a selection, mutation and crossing behavior as the evolution of living beings has done over the years. All this gives rise to the fact that even with similar operators, each metaheuristic algorithm has its own behavior with the purpose of exploring the search space for the problems and the exploitation of the results obtained when carrying out the evaluations of the objective function.

5.5.1 Basic Components in the Proposal of the Metaheuristic Algorithm

Metaheuristics usually focuses on using the best solution obtained so far [8] despite the fact that this method offers ideal results for various problems, it also presents drawbacks that must be addressed. One of these drawbacks is observed when the algorithm focuses only on the best solution obtained, gradually ending the diversity in the solutions since the particles become more and more concentrated in a specific point within the search space. This approach can bring with it other consequences, such as premature convergence, which would lead the algorithm to stagnate in a solution considered to be locally optimal and thus ignore the rest of the search space where there could be better quality solutions compared to the one found.

Another consequence is the creation of a repetitive cycle with the feedback obtained from the particles in the search space, which leads the particles to obtain very similar solutions and thus limits the capacity of the algorithm to be able to create new solutions and the possibility that the new solutions are more effective. This does not mean that the action of using the best solution is a bad strategy for metaheuristics since it can give optimal solutions for optimization problems; however, it is necessary to handle it with caution and not leave out the individual strategy of the particle, that is, not forget the information that it obtained from the search space individually.

With all this information, the goal is to achieve a metaheuristic algorithm that makes use not only of the best solution but also of the information that the search particles have, thus achieving a balance between individual learning and social learning. In order to achieve this, it is essential to apply two different strategies: the individual strategy, where the particle makes use of the information obtained by its own means and performs an independent search within the search space, and the collective strategy, where the particle will make use of the information obtained by the rest of the population by sharing information from the search space through the best solution obtained. In addition, the search space is considered a learning problem where each of the individuals must adjust to determine how frequently each of the mentioned strategies will be used [9]. In order to determine the frequency with which the search particles will work, each of these has its counter with a different stopping criterion, with which we seek to have a variety of strategies between the particles since some will have a higher limit, resulting in greater diversity in the search space.

In Fig. 5.3, we can observe the behavior of this algorithm, where at the beginning, each of the search particles has its own strategy (represented as the area surrounding each particle), which leads to the particles searching for optimal solutions in the vicinity by means of the individual strategy. Once they have completed the limit of their counters, they make use of the collective strategy and use the information of the best particle to move to its position. However, because the counters of each of the particles are different, some particles move before others, so some do not move from their position, allowing for greater diversity in the results.

The final objective of this algorithm is to obtain an optimal solution for the optimization problems, as can be seen in the following equation:

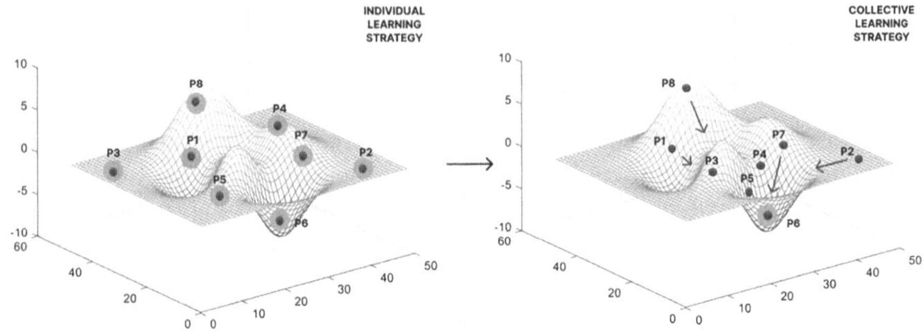

Fig. 5.3 Representation of the use of learning strategies in the algorithm

$$\begin{array}{l} \textit{Minimize} \quad f(x) \quad x = (x_1, x_2, x_3, \ldots, x_4) \in R^d \\ \textit{Subject to} \quad x \in S \end{array} \tag{5.1}$$

In Eq. 5.1, we can see that R is a non-linear function with d dimensions, where d is the search space. This search space is defined by R^d. Similarly, it is restricted by an upper bound u_i and a lower bound l_i where i is each dimension from 1 to d.

5.5.2 Setting Initial Parameters and Initialization

In the first stage, the initial position of each particle within the search space must be established; in the second stage, a standard deviation is assigned to each particle; and finally, a random and different value is assigned to each particle's counter.

For the first stage, what must be done is to start a group of N number of particles which will be called population P, thus being:

$$P = \{x_1, x_2, x_3, \ldots, x_N\} \tag{5.2}$$

In Eq. 5.2, each particle x_i represents a decision variable that, in turn, represents a solution to the optimization problem. The value of each of the decision variables x_i is determined using a uniform distribution between the lower limit (l_i) and the upper limit (u_i) mentioned above; this value is determined through the following equation:

$$x_i = l_i + rand\ (0{,}1) * (u_i - l_i) \tag{5.3}$$

In the second stage, the standard deviation $\{\sigma_1, \sigma_2, \sigma_3, \ldots, \sigma_N\}$ of each particle is fixed. Each standard deviation σ_i of each particle x_i covers a series of d elements, where each element $\sigma_{i,j}$ is randomly initialized with a uniform value between 0 and 1. This standard deviation is very important for the functioning of the algorithm since it is a fundamental part of the search particles' ability to carry out their individual strategies.

Finally, in the third stage, each particle x_i is assigned to its own counter c_i. The value of this counter is fundamental since it determines the frequency with which the particle will use the individual strategy or the collective strategy. Each counter is assigned a random number that ranges from 2 to M, where M represents the total number of iterations (*Niter*) between two to be performed in the algorithm.

5.5.3 Counter Management and Updating

Nowadays, there are metaheuristic algorithms that prefer to focus on the best solution obtained collectively, leaving aside and ignoring the possible individual strategies that the particles in the group have [10, 11]. As mentioned above, although this type of strategy gives optimal results, it can lead to complications, such as premature convergence in the process of finding solutions to the problem.

In this proposed algorithm, each particle x_i has two options of learning strategies: the individual learning strategy (ILS) and the collective learning strategy (CLS). The use of the word learning is due to the fact that the proposed algorithm considers the search strategies as learning problems, and because of this, the particle must adjust and balance the use of both strategies. This balance is obtained thanks to the counters c_i of each particle x_i. In other words, the counter value sets when ILS will be implemented and also when CLS will be implemented for each of the particles in the search space. Therefore, if we assign a different value V to each of the counters, this value will indicate when the particle uses the CLS strategy; that is, the particle will use the collective CLS strategy every V iteration, while the rest of the iterations will be working with the individual ILS strategy. This can be represented as follows:

$$x_i(k+1) = \begin{cases} x_i(k) + R * (x_{BEST} - x_i(k)) & CLS \quad every\ V\ iterations \\ x_i(k) + N(0, \sigma_i) & ILS \quad otherwise \end{cases} \tag{5.4}$$

This mechanism idea (Eq. 5.4) creates a systematic alternative between the two strategies mentioned above, ensuring that work and participation in the resolution of optimization problems both collaboratively as well as individually. In addition, the value V acts as a regulator because it determines the frequency with which the particle or search agent changes its strategies by making use of the information of the rest of the population or by using its knowledge and strategy for exploration in the search space.

As mentioned in the previous subchapter, each counter is assigned a random value within the range of 2 to M. This range is intended to ensure that there can only be one strategy change from the second iteration, thus giving the possibility that all particles have their own start. Thus, the best one among the entire population can be determined for future collective strategies. In addition, if a counter is assigned the maximum possible value, this particle will apply the CLS collective strategy at least once. With the help of this mechanism, we can promote a balance between strategies and thus ensure the exploration of the patterns of the particles, improving the adaptability and efficiency of the algorithm to solve the various problems that exist and will exist in the area of optimization.

5.5.4 Implementation of the Individual Learning Strategy (ILS)

The individual learning strategy (ILS) involves the particle independently exploring the search space, relying and trusting only on its own calculations and decisions. This aspect is of utmost importance because it allows for unique and diverse exploration, where new areas in the search space can be observed. The integration of this strategy is vital for an effective search process, as it promotes independent discovery and individual thinking and avoids uniformity.

The ILS strategy for each agent $x_i(k)$ is focused on generating new solutions or positions by exploring the areas surrounding their current positions. This process involves perturbing the current position of the particle x_i, considering the standard deviation σ_i assigned to it during the initialization stage. Consequently, the new position of the particle $x_i(k+1)$ is calculated randomly considering a modification in its position as can be observed in the following equation:

$$ILS: \quad x_i(k+1) = x_i(k) + N(0, \sigma_i) \tag{5.5}$$

The standard deviation determines the variation of the new position. This approach allows the particles in the population $\{x_1, x_2, x_3, \ldots, x_N\}$ to effectively and systematically try to search different regions in the search space, thereby increasing the chances of finding an optimal solution or one close to the optimal solution.

Equation 5.5 represents a mutation of the Gaussian variance similar to the scheme presented in various metaheuristic methods [12]. The Gaussian variance moves the particle in its zone, that is, its vicinity, to its current position within the search space. This is achieved by modifying the particle's variance by a random number, which determines how much it will move to the agent's new position with respect to the previous position.

In Fig. 5.4, we can observe how each particle has its own regions within the search space, represented by the circles of lower density surrounding the particle. These are the possible spaces where the particle can move with the help of the modification in its Gaussian variances, giving the possibility of following its vicinity and thus maintaining its individual strategies.

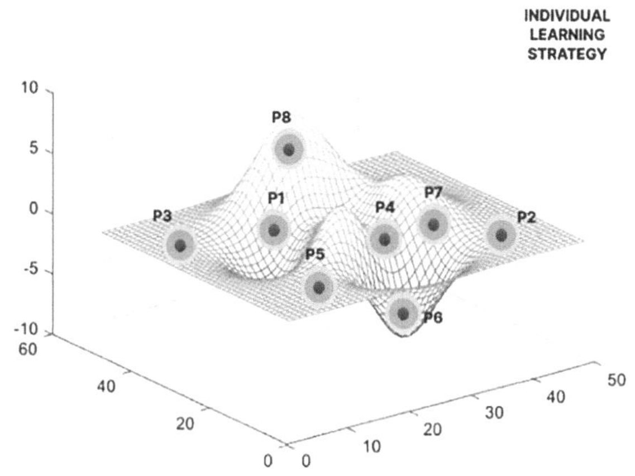

Fig. 5.4 Visual representation of the particles and their individual regions

5.5.5 *Implementation of the Collective Learning Strategy (CLS)*

The collective learning strategy (CLS) is characterized by the particle using the information collected by the rest of the particles in the population. This collective intelligence guides the search of the particle, allowing it to benefit from the experiences and discoveries shared by the rest of the population, regardless of whether these are places with optimal solutions or not, since any information regarding the search space is important for the particle. The integration of this strategy is vital for an effective search process since the collective strategy improves the search by taking advantage of the knowledge of the population, thus achieving an improved and more complete exploration of the different scenarios that involve the resolution of optimization problems.

The CLS strategy for each particle considers the collective knowledge acquired by the group of particles that make up the population $\{x_1, x_2, x_3, \ldots, x_N\}$ to. This strategy focuses particularly on the best solution x_{BEST} found by any other particle in the population up to that moment, based on the criteria of the objective function. This position does not mean that it is the best solution to the problem, but it is considered the best solution obtained so far and can be replaced by the solution of another particle as long as it is better. This optimal position serves as an indicator to identify promising areas in the search space. In other words, it guides the particle x_i to concentrate its search on regions that show potential based on the collective achievements of the group. The best position, obtained from the best solution, acts as a predictor that highlights areas that might yield favorable results, and this plays an important role in the particle's strategy to navigate and explore the search space more effectively. With this CLS strategy, the new position $x_i(k+1)$ of the particle is calculated as follows:

$$CLS: \quad x_i(k+1) = x_i(k) + R * (x_{BEST} - x_i(k)) \tag{5.6}$$

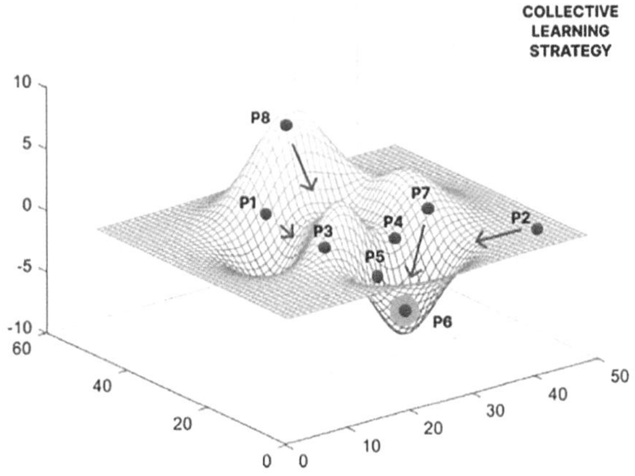

Fig. 5.5 Visual representation of the particles following the best solution

In Eq. 5.6, R represents a random number uniformly distributed between 0 and 1. By doing so, the CLS strategy effectively attracts the particle position x_i to the best position x_{BEST} found so far in the search space. Therefore, the particle switches from a local search to a more promising area, enhancing the quality of the solution. This mechanism is a strategic modification since it involves moving the particle away from less promising areas and aligning its exploration process with better areas that have already demonstrated good potential in terms of the quality of the solutions.

In Fig. 5.5, we can observe how the particles move following the collective strategy when their counters manage to reach the limit, thus following the best particle found within the population and using its information to move to a more promising region within the search space. In short, this strategy is a form of guided exploration in which the search path is optimized to focus on parts of the search space that are more likely to produce superior optimal solutions, i.e., of better quality. This strategy generally improves the efficiency and effectiveness of the search process, thus making it more oriented towards optimal results.

5.5.6 Computational Implementation of the Proposal

This subchapter, accompanied by the pseudocode Algorithm 1, can observe the computational process of the proposed algorithm. This makes it easier to understand. This algorithm requires two parameter inputs (line 1) from the user: the number of agents N to be used and the maximum number of iterations $Niter$ to be used throughout the algorithm optimization process. To begin, the algorithm goes through an initialization phase (lines 2, 3 and 4), in which the initial positions of the search particles are established $\{x_1, x_2, x_3, \ldots, x_N\}$, the standard deviations of each of these particles $\{\sigma_1, \sigma_2, \sigma_3, \ldots, \sigma_N\}$ and the value of the counters $\{c_1, c_2, c_3, \ldots, c_N\}$ of each one with which the control between the learning strategies to be used will be carried out. Following the initialization phase, the algorithm enters a repetitive cycle (line 5), in which it will continue until the maximum number of iterations *Niter* previously established by the user are met.

In each of these cycles, the algorithm examines the value of each counter c_i of each particle x_i (line 7) and this counter, as mentioned above, determines the strategy to be used by the particle. If the counter value is not equal to the set limit ($c_i \neq V$), then the ILS will be executed (lines 8 and 9), while if the counter value is equal to the set limit ($c_i = V$), then the CLS will be executed (lines 10, 11 and 12). It is worth mentioning that this operating mechanism allows a resolutive balance between individual exploration and collective learning, thus improving the effectiveness of the algorithm when moving through the search space.

Algorithm 1 Pseudocode of the Proposed Algorithm

1. Input of data: $N, Niter, k = 0$
2. Population Initialization → $\{x_1, x_2, x_3, \ldots, x_N\}$
3. Initialization of the Standard Deviation → $\{\sigma_1, \sigma_2, \sigma_3, \ldots, \sigma_N\}$
4. Initialization of Counters → $\{c_1, c_2, c_3, \ldots, c_N\}$
5. **while** $k \leq Niter$ **do**
6.   **for** each particle x_i who is in $\{x_1, x_2, x_3, \ldots, x_N\}$
7.   Compare the counter c_i value with respect to V
8.   **if** $c_i \neq V$
9. $x_i(k+1) = x_i(k) + N(0, \sigma_i)$
10.   **else if** $c_i = V$
11.   Find the best solution → x_{BEST}
12. $x_i(k+1) = x_i(k) + R * (x_{BEST} - x_i(k))$
13.   **end**
14.   **end**
15. **end**
16. Data output: x_{BEST}

To conclude, the following Fig. 5.6 shows a flow chart explaining the basic behavior of this metaheuristic algorithm thoroughly.

5.6 Evaluation, Testing and Analysis of the Algorithm

This subchapter introduces several sets of experiments that were conducted to evaluate the performance of the proposed algorithm, which will be named "Individual and Collective Search Strategy" (ICSS). The experiments conducted comprehensively evaluate the effectiveness and robustness of the proposed method through detailed numerical and statistical analysis. For easier analysis, a set of 20 test functions was used, including unimodal, multimodal and hybrid functions, which are functions constituted by the combination of two or more unimodal or multimodal functions [13]. All the mathematical formulas for the functions used during the experimentation stage are detailed in Table 5.13 in the appendix of this chapter. Under this concept, the nomenclature $f(x^*)$ represents the optimal value of each of the functions

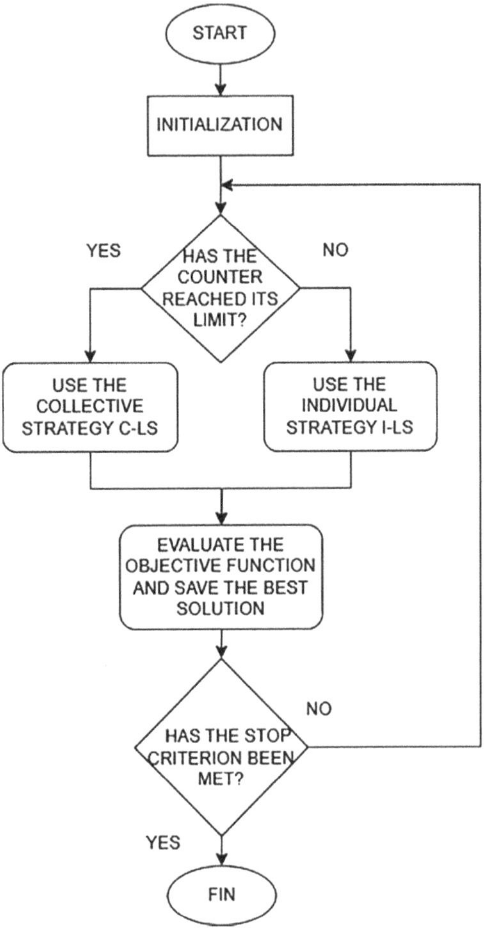

Fig. 5.6 ICSS algorithm flowchart

when they are evaluated in the function x^*, while $\boldsymbol{S}$ indicates the search space defined by its upper and lower limits.

In order to ensure a thorough comparison of the performance of the proposed algorithm, a selection of several metaheuristic algorithms known in the literature has been made to be used as a point of comparison in the analysis. Some authors consider the selected algorithms to be optimization techniques considered both classical and modern. The list of selected algorithms is the following: Gray Wolf Optimizer (GWO) [14], Differential Evolution (DE) [15], Particle Swarm Optimization (PSO) [16], Crow Search Algorithm (CSA) [17], Moth-Flame Optimization Algorithm (MFO) [18] and Genetic Algorithm (GA) [19, 20], thus achieving a total of six algorithms with which the comparisons were made. In order to maintain a clear consistency when making the comparisons, each of the algorithms was configured according to the specifications given by the original authors. The following Table 5.3 shows the

Table 5.3 Setting the parameters for each metaheuristic algorithm used in the comparison

Algorithm	Parameters	References
GWO	The population size is $pop = 50$	[14]
DE	The variant considered is DE/rand/bin, where crossover probability $CR = 0.5$ and differential weight $dw = 0.2$	[15]
PSO	The parameters are set to $c1 = 2$ and $c2 = 2$ with linearly decreasing weight factor from 0.9 to 0.2	[16]
CSA	The awareness probability is set as $AP = 0.1$ and the flight length $fl = 2$	[17]
MFO	The population size is $pop = 50$	[18]
GA	The crossover probability $cr = 0.7$, the mutation probability is set at 0.3, and the mutation rate $mu = 0.1$	[19, 20]

specifications of the parameters used in each of the metaheuristic algorithms used in the experiment.

This experimental setup allows for a precise evaluation of each algorithm's performance, delimiting their respective strengths and limitations. In this way, the comparative approach provides an exhaustive analysis, emphasizing the proposed algorithm's effectiveness within the various metaheuristic strategies existing in the literature.

5.6.1 Performance Analysis in the Functions

In the comparative analysis of algorithms, the focus is mainly on the fitness values obtained (mostly known in the literature as fitness) generated by the different algorithms, with the aim of minimizing these values during the optimization process. A maximum number of iterations was already predefined in the previous subchapters, set at $Niter = 1000$, which has the function of being the stopping criterion and thus regulating the optimization process. Also, the evaluation examines the performance of the algorithms through different work scales; specifically, in this case, we talk about 30 and 100 dimensions in order to evaluate the scalability of the proposed method.

For each of the algorithms, the population size is set at 50 individuals, thus defining the number of candidate solutions generated and evaluated in each of the iterations. Likewise, due to the stochastic nature of these algorithms, each of the test functions is subjected to 30 independent executions to ensure an exhaustive evaluation because multiple executions allow a deeper understanding of how randomness and variability affect the performance of the algorithm, providing valuable data for the analysis and statistical interpretation of these algorithms.

The results of each of the algorithms are detailed in the following Tables 5.4 and 5.5, which provide essential numerical metrics to evaluate the performance of each of the algorithms. Among the metrics, we can find the best fitness value (f_{BEST}), the mean fitness values $\left(\overline{f}\right)$ and the standard deviation $\left(\sigma_f\right)$. The best fitness value represents

Table 5.4 Results of minimization of the benchmark functions in 30 dimensions

		GWO	DE	PSO	CSA	MFO	GA	ICSS
f_1	f_{Best}	1.21E-14	2.86E+00	5.62E-06	1.72E+01	7.50E+00	2.24E-02	**8.88E-16**
	$\bar{f}$	1.42E-14	2.83E+00	2.52E-06	1.72E+01	1.91E+00	2.11E-02	8.88E-16
	σ_f	3.18E-15	3.34E-01	1.10E-05	4.16E-01	8.41E+00	9.20E-03	0.00E+00
f_2	f_{Best}	**6.67E-01**	2.74E+01	1.32E+02	1.57E+05	4.26E+04	3.07E+00	6.69E-01
	$\bar{f}$	6.67E-01	2.66E+01	9.04E+01	1.49E+05	4.50E+01	2.88E+00	6.76E-01
	σ_f	2.33E-06	7.00E+00	1.41E+02	3.67E+04	9.03E+04	1.59E+00	4.99E-03
f_3	f_{Best}	7.89E-04	1.05E+00	9.02E-03	1.75E+02	1.51E+01	4.59E-02	**0.00E+00**
	$\bar{f}$	0.00E+00	1.05E+00	7.40E-03	1.79E+02	1.24E-02	4.45E-02	0.00E+00
	σ_f	3.15E-03	1.07E-02	1.09E-02	2.29E+01	3.43E+01	2.90E-02	0.00E+00
f_4	f_{Best}	1.96E-165	1.46E-07	**0.00E+00**	1.19E-01	3.86E-02	1.26E-15	**0.00E+00**
	$\bar{f}$	1.45E-167	1.10E-07	0.00E+00	1.22E-01	2.60E-16	4.22E-16	0..00E+00
	σ_f	0.00E+00	9.71E-08	0.00E+00	4.28E-02	2.12E-01	2.34E-15	0.00E+00
f_5	f_{Best}	9.78E-01	7.95E-01	1.51E-01	5.72E+01	2.69E+01	**1.41E-04**	1.44E-01
	$\bar{f}$	9.98E-01	7.79E-01	4.58E-11	5.77E+01	2.63E+01	8.79E-05	3.23E-01
	σ_f	2.23E-01	1.25E-01	2.48E-01	7.14E+00	9.02E+00	1.48E-04	6.54E-02
f_6	f_{Best}	1.29E-11	**0.00E+00**	4.44E-30	8.63E-02	**0.00E+00**	5.20E-13	**0.00E+00**
	$\bar{f}$	8.78E-12	0.00E+00	0.00E+00	8.67E-02	0.00E+00	2.47E-13	3.36E-04
	σ_f	1.24E-11	0.00E+00	2.09E-29	1.98E-02	0.00E+00	6.19E-13	1.07E-03
f_7	f_{Best}	**0.00E+00**	2.32E-66	**0.00E+00**	1.31E+02	4.82E-200	**0.00E+00**	**0.00E+00**
	$\bar{f}$	0.00E+00	3.57E-68	0.00E+00	2.07E+01	1.59E-221	0.00E+00	0.00E+00
	σ_f	0.00E+00	9.01E-66	0.00E+00	2.74E+02	0.00E+00	0.00E+00	0.00E+00
f_8	f_{Best}	9.58E+04	1.52E+05	9.92E+01	7.15E+07	1.37E+07	8.40E+01	**4.81E-01**
	$\bar{f}$	9.39E+04	1.53E+05	9.94E+01	6.50E+07	9.36E+01	8.38E+01	1.43E+00
	σ_f	7.30E+04	3.68E+04	5.29E+00	2.26E+07	7.49E+07	4.04E+00	5.52E-01
f_9	f_{Best}	**3.00E+01**	**3.00E+01**	1.28E+02	5.86E+01	3.54E+01	**3.00E+01**	**3.00E+01**
	$\bar{f}$	3.00E+01	3.00E+01	1.27E+02	5.95E+01	3.50E+01	3.00E+01	3.12E+01
	σ_f	0.00E+00	0.00E+00	4.08E+00	2.84E+00	5.54E+00	0.00E+00	6.99E-01
f_{10}	f_{Best}	3.55E-06	3.89E+02	1.81E+02	1.89E+03	7.76E+02	1.97E-01	**0.00E+00**
	$\bar{f}$	1.96E-06	4.00E+02	1.55E+02	1.88E+03	2.84E+02	1.91E-01	0.00E+00
	σ_f	4.36E-06	6.65E+01	1.12E+02	3.79E+02	9.78E+02	5.79E-02	0.00E+00
f_{11}	f_{Best}	9.44E+02	2.26E+05	**2.08E-05**	1.99E+10	4.17E+09	1.35E+01	7.53E+02
	$\bar{f}$	8.71E+02	2.22E+05	1.74E-05	2.01E+10	3.29E-01	7.97E+00	9.94E+02
	σ_f	4.94E+02	7.44E+04	2.03E-05	4.75E+09	1.59E+10	1.63E+01	1.42E+02
f_{12}	f_{Best}	8.98E+00	1.19E+01	1.07E+01	2.45E+01	1.39E+01	9.37E+00	**7.92E+00**
	$\bar{f}$	8.98E+00	1.20E+01	1.08E+01	2.46E+01	1.12E+01	9.45E+00	1.11E+01
	σ_f	3.75E-01	5.87E-01	7.62E-01	2.99E+00	6.50E+00	5.12E-01	7.86E-01
f_{13}	f_{Best}	1.06E-01	4.16E+01	5.06E+01	2.60E+02	1.54E+02	1.38E-01	**0.00E+00**

(continued)

Table 5.4 (continued)

		GWO	DE	PSO	CSA	MFO	GA	ICSS
	$\bar{f}$	0.00E+00	4.20E+01	4.63E+01	2.60E+02	1.54E+02	2.56E-02	0.00E+00
	σ_f	5.81E-01	4.07E+00	1.78E+01	1.39E+01	3.52E+01	3.13E-01	0.00E+00
f_{14}	f_{Best}	**2.65E+01**	3.73E+02	2.63E+04	2.07E+05	6.70E+04	6.46E+01	2.84E+01
	$\bar{f}$	2.62E+01	3.75E+02	2.52E+03	1.95E+05	5.55E+04	7.28E+01	2.85E+01
	σ_f	6.41E-01	5.91E+01	3.77E+04	5.87E+04	6.76E+04	4.41E+01	4.70E-02
f_{15}	f_{Best}	7.84E-40	4.89E+00	2.33E+02	5.20E+29	3.90E+02	6.98E-02	**0.00E+00**
	$\bar{f}$	3.84E-40	4.65E+00	2.28E+02	2.20E+29	4.00E+02	5.63E-02	0.00E+00
	σ_f	1.20E-39	8.85E-01	9.88E+01	1.13E+30	1.83E+02	4.88E-02	0.00E+00
f_{16}	f_{Best}	5.79E-70	2.51E+01	2.43E+03	1.19E+05	1.26E+04	1.08E-01	**0.00E+00**
	$\bar{f}$	9.25E-71	2.48E+01	3.96E-10	1.16E+05	4.29E+03	7.70E-02	0.00E+00
	σ_f	1.32E-69	5.16E+00	5.13E+03	1.31E+04	2.35E+04	8.57E-02	0.00E+00
f_{17}	f_{Best}	4.76E-73	1.49E-02	6.02E-13	5.15E+01	1.75E+00	6.10E-05	**0.00E+00**
	$\bar{f}$	7.27E-74	1.42E-02	1.98E-13	5.34E+01	6.90E-08	4.69E-05	0.00E+00
	σ_f	1.09E-72	3.18E-03	1.09E-12	7.24E+00	6.65E+00	5.78E-05	0.00E+00
f_{18}	f_{Best}	6.21E-229	1.71E-10	4.63E-25	8.06E-03	8.86E-21	3.03E-07	**0.00E+00**
	$\bar{f}$	7.65E-240	1.18E-10	1.94E-28	7.02E-03	4.14E-22	1.15E-13	0.00E+00
	σ_f	0.00E+00	1.56E-10	2.49E-24	4.70E-03	4.15E-20	1.65E-06	0.00E+00
f_{19}	f_{Best}	8.60E-41	7.92E+00	1.00E+03	2.03E+04	2.28E+04	4.17E-02	**0.00E+00**
	$\bar{f}$	5.37E-41	7.56E+00	4.95E-07	2.12E+04	1.51E+04	3.16E-02	0.00E+00
	σ_f	9.57E-41	1.76E+00	3.06E+03	2.84E+03	2.37E+04	3.19E-02	0.00E+00
f_{20}	f_{Best}	**2.90E+01**	8.07E+01	6.61E+01	7.46E+02	1.58E+02	2.91E+01	**2.90E+01**
	$\bar{f}$	2.90E+01	8.22E+01	6.91E+01	7.47E+02	1.08E+02	2.91E+01	2.89E+01
	σ_f	5.21E-05	1.18E+01	1.61E+01	5.15E+01	1.41E+02	1.96E-01	7.94E-04

the lowest fitness value (because the optimization problems were taken as minimization problems) achieved by each of the methods, thus reflecting the effectiveness of the solutions generated by the methods. Under this same context, the lower the value, the higher the performance. The mean fitness value is the average value calculated over all the executions of each method, thus offering a general perspective of the algorithm's performance and its consistency in the functions. Finally, the standard deviation quantifies the variability in the fitness values throughout the 30 executions of each method. That said, a lower standard deviation means greater consistency and more reliable performance. To facilitate the identification of the most efficient algorithms, because there are functions where more than one algorithm had optimal results, achieving the best possible solution in the search space, the entries that show better performance are highlighted in bold.

Table 5.4 details the numerical results of the experiments performed in 30 dimensions. The proposed ICSS algorithm demonstrates a superior fitness value on twelve functions ($f_1, f_3, f_4, f_8, f_{10}, f_{12}, f_{13}, f_{16}, f_{17}, f_{18}, f_{19}$, and f_{20}) compared to the other

Table 5.5 Results of minimization of the benchmark functions in 100 dimensions

		GWO	DE	PSO	CSA	MFO	GA	ICSS
f_1	f_{Best}	6.79E-14	1.94E+01	7.43E+00	1.88E+01	1.92E+01	2.47E-01	**8.88E-16**
	$\bar{f}$	6.75E-14	1.94E+01	5.70E+00	1.88E+01	1.92E+01	1.99E-01	1.40E-10
	σ_f	4.79E-15	9.83E-02	3.09E+00	1.58E-01	3.46E-01	1.15E-01	7.67E-10
f_2	f_{Best}	**6.67E-01**	1.01E+07	3.22E+05	5.42E+06	1.60E+06	8.13E+01	**6.67E-01**
	$\bar{f}$	6.67E-01	1.02E+07	1.99E+05	5.48E+06	1.18E+06	7.79E+01	9.83E-01
	σ_f	6.45E-07	7.96E+05	3.74E+05	5.32E+05	1.43E+06	1.46E+01	6.12E-02
f_3	f_{Best}	1.37E-03	1.17E+03	2.33E+01	9.77E+02	2.32E+02	9.76E-01	**0.00E+00**
	$\bar{f}$	0.00E+00	1.17E+03	2.29E+00	9.71E+02	2.17E+02	9.92E-01	3.83E+00
	σ_f	4.30E-03	8.17E+01	3.85E+01	4.13E+01	1.15E+02	8.96E-02	1.87E+01
f_4	f_{Best}	3.49E-80	9.56E+00	**0.00E+00**	2.03E+00	5.29E-01	5.88E-10	**0.00E+00**
	$\bar{f}$	5.68E-83	9.90E+00	0.00E+00	2.09E+00	1.39E-01	4.34E-10	0.00E+00
	σ_f	1.83E-79	1.23E+00	0.00E+00	4.26E-01	8.50E-01	4.67E-10	0.00E+00
f_5	f_{Best}	6.18E+00	6.62E+02	6.26E+01	3.59E+02	1.75E+02	**3.43E-02**	5.87E+00
	$\bar{f}$	6.12E+00	6.58E+02	6.13E+01	3.62E+02	1.71E+02	2.96E-02	8.66E+00
	σ_f	2.93E-01	4.16E+01	1.49E+01	2.67E+01	3.68E+01	1.92E-02	6.37E-01
f_6	f_{Best}	1.89E-12	**0.00E+00**	7.75E-05	3.37E-01	3.64E-04	2.67E-09	**0.00E+00**
	$\bar{f}$	5.30E-15	0.00E+00	0.00E+00	3.42E-01	0.00E+00	2.10E-09	1.28E-46
	σ_f	6.95E-12	0.00E+00	4.25E-04	4.00E-02	1.36E-03	1.66E-09	7.02E-46
f_7	f_{Best}	**0.00E+00**	3.04E+05	**0.00E+00**	3.74E+20	7.44E-199	**0.00E+00**	**0.00E+00**
	$\bar{f}$	0.00E+00	1.05E+03	0.00E+00	1.71E+19	3.07E-215	0.00E+00	0.00E+00
	σ_f	0.00E+00	1.08E+06	0.00E+00	1.00E+21	0.00E+00	0.00E+00	0.00E+00
f_8	f_{Best}	2.20E+06	2.59E+09	1.72E+06	8.59E+08	1.77E+08	5.96E+02	**9.91E+00**
	$\bar{f}$	2.13E+06	2.60E+09	1.54E+06	8.73E+08	6.35E+07	5.75E+02	1.40E+01
	σ_f	2.89E+05	2.66E+08	9.34E+05	1.07E+08	2.51E+08	7.78E+01	3.46E+00
f_9	f_{Best}	3.05E+01	1.61E+02	4.60E+01	1.61E+02	7.76E+01	**3.00E+01**	5.60E+01
	$\bar{f}$	3.00E+01	1.60E+02	4.55E+01	1.62E+02	7.70E+01	3.00E+01	6.18E+01
	σ_f	8.19E-01	4.85E+00	7.32E+00	4.73E+00	1.33E+01	0.00E+00	3.43E+00
f_{10}	f_{Best}	7.63E-06	4.63E+04	5.37E+03	1.92E+04	1.12E+04	8.99E+00	**0.00E+00**
	$\bar{f}$	4.36E-06	4.65E+04	4.41E+03	1.95E+04	1.10E+04	8.66E+00	8.27E-252
	σ_f	8.17E-06	4.61E+03	2.94E+03	2.53E+03	5.70E+03	3.06E+00	0.00E+00
f_{11}	f_{Best}	1.32E+05	4.49E+11	9.65E+07	1.85E+11	3.95E+10	**1.31E+04**	1.01E+05
	$\bar{f}$	1.31E+05	4.53E+11	8.30E+07	1.88E+11	2.38E+10	1.17E+04	4.81E+07
	σ_f	1.83E+04	3.80E+10	6.43E+07	1.97E+10	3.77E+10	5.75E+03	2.31E+08
f_{12}	f_{Best}	3.93E+01	7.03E+02	7.42E+01	4.02E+02	2.49E+02	4.88E+01	**3.77E+01**
	$\bar{f}$	3.94E+01	6.99E+02	6.30E+01	4.12E+02	2.27E+02	4.90E+01	4.44E+01
	σ_f	9.78E-01	5.41E+01	2.54E+01	4.97E+01	1.07E+02	2.93E+00	1.59E+00
f_{13}	f_{Best}	9.47E-14	1.12E+03	4.71E+02	1.13E+03	7.12E+02	7.55E+01	**0.00E+00**

(continued)

Table 5.5 (continued)

		GWO	DE	PSO	CSA	MFO	GA	ICSS
	$\bar{f}$	1.14E-13	1.13E+03	4.58E+02	1.12E+03	7.12E+02	7.31E+01	0.00E+00
	σ_f	5.24E-14	2.17E+01	6.58E+01	2.82E+01	8.63E+01	1.25E+01	0.00E+00
f_{14}	f_{Best}	**9.71E+01**	1.15E+06	1.03E+06	2.32E+06	8.05E+05	3.83E+02	9.82E+01
	$\bar{f}$	9.69E+01	1.16E+06	1.10E+06	2.40E+06	7.16E+05	3.81E+02	9.88E+01
	σ_f	7.37E-01	9.47E+04	2.82E+05	3.19E+05	5.04E+05	6.27E+01	2.34E-01
f_{15}	f_{Best}	9.60E-20	5.72E+102	2.17E+03	1.30E+118	2.32E+03	5.77E+00	**0.00E+00**
	$\bar{f}$	8.12E-20	1.93E+100	2.13E+03	1.07E+117	2.32E+03	5.64E+00	5.91E-13
	σ_f	5.90E-20	2.32E+103	2.36E+02	3.51E+118	4.40E+02	9.91E-01	3.24E-12
f_{16}	f_{Best}	2.31E-33	1.89E+06	2.20E+05	2.17E+06	7.00E+05	8.01E+01	**0.00E+00**
	$\bar{f}$	1.57E-33	1.90E+06	1.67E+05	2.19E+06	5.57E+05	7.86E+01	7.21E+03
	σ_f	2.18E-33	1.09E+05	1.59E+05	1.37E+05	4.56E+05	1.92E+01	3.95E+04
f_{17}	f_{Best}	4.61E-37	3.45E+02	7.36E+00	2.85E+02	5.81E+01	1.14E-02	**0.00E+00**
	$\bar{f}$	4.00E-37	3.47E+02	3.38E-01	2.85E+02	5.03E+01	1.07E-02	7.70E-18
	σ_f	3.69E-37	1.86E+01	1.37E+01	1.15E+01	3.06E+01	3.32E-03	3.88E-17
f_{18}	f_{Best}	3.41E-182	1.16E+00	1.12E-07	1.09E-02	1.19E-04	2.90E-07	**0.00E+00**
	$\bar{f}$	2.24E-196	1.15E+00	1.26E-08	1.01E-02	1.16E-05	5.84E-09	0.00E+00
	σ_f	0.00E+00	3.31E-01	2.38E-07	5.15E-03	2.89E-04	5.40E-07	0.00E+00
f_{19}	f_{Best}	2.83E-20	2.43E+05	6.35E+04	5.57E+18	1.51E+05	6.58E+00	**0.00E+00**
	$\bar{f}$	2.39E-20	2.38E+05	6.06E+04	2.62E+16	1.45E+05	6.50E+00	9.54E-02
	σ_f	1.62E-20	2.30E+04	3.47E+04	2.04E+19	6.07E+04	1.66E+00	5.23E-01
f_{20}	f_{Best}	**9.90E+01**	5.14E+03	6.62E+02	3.62E+03	1.69E+03	1.03E+02	**9.90E+01**
	$\bar{f}$	9.90E+01	5.18E+03	6.01E+02	3.63E+03	1.59E+03	1.03E+02	9.90E+01
	σ_f	3.95E-04	2.24E+02	1.99E+02	1.23E+02	5.10E+02	9.53E-01	4.46E-03

metaheuristic algorithms. These results indicate the efficiency of ICSS in identifying optimal solutions, especially in those problems with multiple local optima. ICSS shows comparable performance to GWO on the functions f_2, f_7, f_9, f_{14} and f_{20}. Likewise, in the functions f_7 and f_9, PSO and GA algorithms produce results similar to those of ICSS. We can observe the same with the DE in the functions f_6 and f_9. The MFO algorithm also reflects the ICSS results on the function f_6. These solutions suggest that GWO, PSO and GA often produce results similar to ICSS. The same is true for DE and MFO, only less frequently. Similarly, ICSS shows suboptimal performance on two functions, this one being the f_5 and f_{11}.

These similarities in results with GWO, PSO, and MFO can be attributed to their incorporation of multiple influential solutions in the search process. Similarly, ICSS's high effectiveness is largely due to the coordination of the I-LS and C-LS strategies between its search particles. This combination integrates independent exploration with a bias towards the most promising areas in the search space, as indicated by the best solution obtained so far under the collective information.

On the other hand, Table 5.5 presents the results of the scenario where the functions were subjected to 100 dimensions, thus offering information about the scalability of the ICSS algorithm. The findings from these high-dimensional tests indicate that the proposed method outperforms other methods on most of the benchmark functions even when the dimensions are expanded to 100. ICSS excels in achieving optimal solutions on sixteen functions, highlighting its ability to maintain effectiveness even when the dimensionality of the optimization problems increases. However, it is worth mentioning that ICSS shows suboptimal results on four functions, namely functions f_5, f_9, f_{11} and f_{14}. In these tests, the strongest competitors were GWO and GA. This competitiveness can be attributed to certain similarities between the behavior of the operators used in these two algorithms and the proposed algorithm, particularly the fact that they emphasize collaborative efforts. Furthermore, the results in Table 5.5 demonstrate not only the robust performance of the ICSS algorithm in high-dimensional spaces but also demonstrate the competitive landscape of metaheuristic algorithms in such complex scenarios. Optimizing functions in 100 dimensions is inherently more challenging than doing so in 30 dimensions; that said, due to the increased complexity, the proposed algorithm has demonstrated remarkable efficiency. The key to the effectiveness lies in the unique way each particle utilizes its I-LS strategy and its C-LS strategy, endowing the particles with a distinctive balance in the application of these strategies, resulting in unique search patterns.

Given the stochastic nature of metaheuristic algorithms, their performance may vary between different runs. This variability requires statistical analysis in order to determine whether the different performances observed between the two algorithms are statistically significant or simply the result of chance. In order to address this in the research, the Wilcoxon signed rank test [21–23] has been used. In practice, this test involves comparing the performance results of two algorithms on the same set of benchmark functions. A crucial output of the Wilcoxon test is the p-value, which quantifies the probability that any observed difference in performance is due to randomness. If this p-value falls below a significance threshold (which is set at 0.5 for this study), the differences in performance are considered statistically significant. This tells us that the observed variations in the algorithm's performance are unlikely to be due to random factors, thus providing a more reliable assessment of its comparative effectiveness.

The statistical analysis performed in this study is based on the results shown in Table 5.5, corresponding to the functions in 100 dimensions. The results of the Wilcoxon test are detailed in Table 5.6. This table provides a pairwise comparison between the ICSS algorithm and another of the algorithms mentioned in the experiment, thus allowing the statistical validation of the experiments performed. For easier interpretation, Table 5.6 uses a symbology consisting of the figures ▲, ▼, and ▶ to represent the different results obtained in the Wilcoxon test. The symbol ▲ indicates that the ICSS algorithm outperformed the competing algorithm, while the symbol ▼ indicates that the competing algorithm outperformed the ICSS algorithm. Finally, the symbol ▶ is used when the Wilcoxon test does not find any significant difference between the algorithms.

Table 5.6 Results of Wilcoxon analysis according to the results in 100 dimensions

ICSS VS	GWO	DE	PSO	CSA	MFO	GA
$f_1(x)$	3.12E-11▲	1.72E-12▲	1.72E-12▲	1.72E-12▲	1.72E-12▲	1.72E-12▲
$f_2(x)$	3.23E-01►	2.80E-11▲	2.80E-11▲	2.80E-11▲	2.80E-11▲	2.80E-11▲
$f_3(x)$	2.80E-11▲	6.48E-12▲	1.37E-09▲	6.48E-12▲	8.02E-12▲	2.65E-09▲
$f_4(x)$	1.21E-12▲	1.21E-12▲	0.00E00►	1.21E-12▲	1.21E-12▲	1.21E-12▲
$f_5(x)$	3.82E-10▲	3.02E-11▲	3.02E-11▲	3.02E-11▲	3.02E-11▲	3.02E-11▼
$f_6(x)$	2.11E-11▲	1.27E-05▲	1.24E-04▲	2.11E-11▲	1.20E-01►	2.11E-11▲
$f_7(x)$	0.00E00►	1.21E-12▲	0.00E00►	1.21E-12▲	1.21E-12▲	0.00E00►
$f_8(x)$	3.02E-11▲	3.02E-11▲	3.02E-11▲	3.02E-11▲	3.02E-11▲	3.02E-11▲
$f_9(x)$	1.27E-11▼	2.73E-11▲	9.90E-11▼	2.70E-11▲	2.52E-06▲	1.09E-12▼
$f_{10}(x)$	1.72E-12▲	1.72E-12▲	1.72E-12▲	1.72E-12▲	1.72E-12▲	1.72E-12▲
$f_{11}(x)$	4.73E-01►	3.02E-11▲	5.97E-09▲	3.02E-11▲	3.02E-11▲	3.02E-11▼
$f_{12}(x)$	5.07E-10▲	3.02E-11▲	3.02E-11▲	3.02E-11▲	3.02E-11▲	5.53E-08▲
$f_{13}(x)$	4.67E-10▲	1.21E-12▲	1.21E-12▲	1.21E-12▲	1.21E-12▲	1.21E-12▲
$f_{14}(x)$	4.98E-11▼	3.02E-11▲	3.02E-11▲	3.02E-11▲	3.02E-11▲	3.02E-11▲
$f_{15}(x)$	2.80E-11▲	2.80E-11▲	2.80E-11▲	2.80E-11▲	2.80E-11▲	2.80E-11▲
$f_{16}(x)$	8.93E-06▲	2.63E-11▲	1.56E-10▲	2.63E-11▲	3.22E-11▲	4.94E-10▲
$f_{17}(x)$	9.12E-07▲	2.11E-11▲	2.11E-11▲	2.11E-11▲	2.11E-11▲	2.11E-11▲
$f_{18}(x)$	1.21E-12▲	1.21E-12▲	1.21E-12▲	1.21E-12▲	1.21E-12▲	1.21E-12▲
$f_{19}(x)$	3.99E-04▲	3.02E-11▲	3.02E-11▲	3.02E-11▲	3.02E-11▲	3.02E-11▲
$f_{20}(x)$	4.80E-07▲	3.02E-11▲	3.02E-11▲	3.02E-11▲	3.02E-11▲	3.02E-11▲
Total ▲	15	20	17	20	19	16
Total ►	3	0	2	0	1	1
Total ▼	2	0	1	0	0	3

According to the p-values shown in Table 5.6, the ICSS algorithm demonstrates superiority over the DE and CSA algorithms. It can also be observed that the GWO algorithm shows similar performance to the ICSS algorithm on functions f_2, f_7 and f_{11}. The Wilcoxon test also indicates that there is no significant difference between the ICSS and PSO algorithms on functions f_6 and f_7. On function f_7, the GA demonstrates comparable performance to the ICSS algorithm. However, on functions f_5, f_9, f_{10} and f_{14}, the ICSS algorithm shows suboptimal results compared to the GA and GWO algorithms. An interesting detail is that the MFO algorithm has a good performance on function f_6. Yet the Wilcoxon test does not distinguish any significant difference between it and the ICSS algorithm.

Overall, the Wilcoxon test results attest to the effectiveness of the ICSS algorithm. However, the GWO and GA algorithms show competitive results on the set of benchmark functions; the ICSS algorithm maintains its effectiveness, robustness and scalability in most of the evaluated functions.

5.6.2 Convergence Analysis of Functions

Evaluating the effectiveness of an optimization algorithm involves more than just the quality of the final solution but also how quickly it converges to that solution. Convergence analysis [24] plays a key role in studying the behavior of a metaheuristic algorithm and its progression to an optimal solution in successive iterations. This type of analysis is essential in measuring the performance of a metaheuristic algorithm and its ability to achieve high-quality solutions as the iterations progress consistently. In this subchapter, a convergence analysis is introduced, comparing the ICSS algorithm with other metaheuristics used in the experimentation phase. For an illustrative perspective, Fig. 5.7 shows representative convergence plots based on the 100-dimensional test function results given in Table 5.5. It is worth noting that, as with the performance analysis of the algorithms, the stopping criterion was set at 1000 iterations. Furthermore, each simulation was run 30 times independently for all metaheuristic algorithms, and the convergence data represent the median of these runs.

Analyzing the data shown in Fig. 5.7, the ICSS algorithm demonstrates an advantage: it converges faster than the other algorithms in the study. This efficiency is attributed to its dual-strategy approach. It starts by efficiently analyzing regions

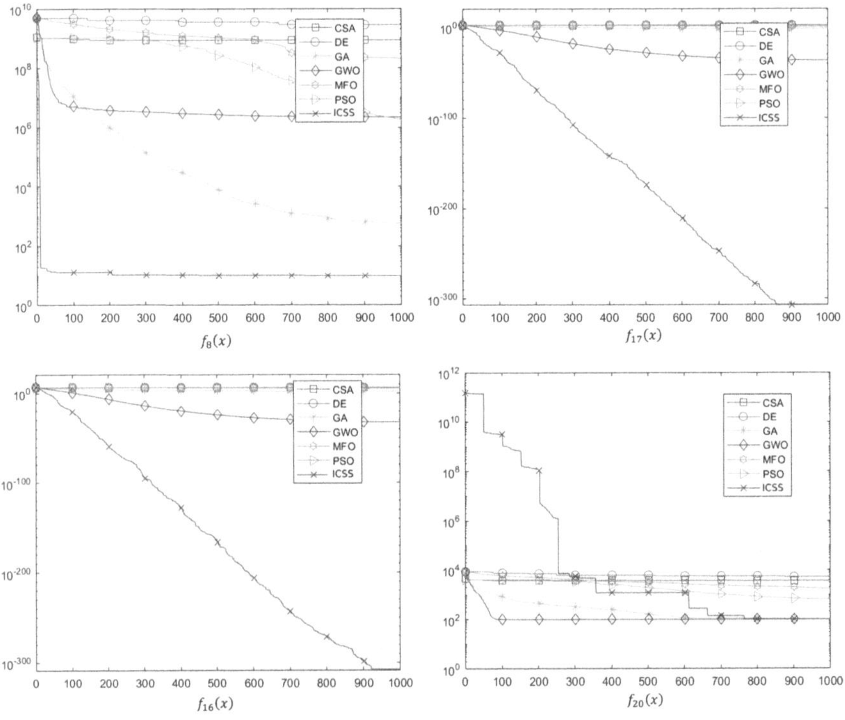

Fig. 5.7 Convergence graphics of the functions f_8, f_{16}, f_{17} and f_{20} in 100 dimensions

of interest by means of the I-LS of each particle, allowing a rapid evaluation of promising areas in the search space. Once one of the particles identifies the best solution, it communicates it to the rest of the population by means of the C-LS. This information triggers a subtle but strategic adaptation in the particles, directing and aligning them toward more promising positions. This mechanism not only speeds up the convergence process but also ensures that the algorithm responds dynamically to the evolution in the search space, thus improving the efficiency of solving optimization problems.

5.6.3 CEC 2017 Experimentation

The above experiments were carried out in order to facilitate the evaluation of the effectiveness and robustness of the ICSS algorithm on conventional benchmark functions, including unimodal, multimodal and hybrid. However, the Congress on Evolutionary Computation (CEC) [25, 26] is one of the largest conferences within Evolutionary Computation, where different tests for metaheuristic algorithms have been developed to test their capabilities, such as the CEC 2017 benchmark functions. So, to complement the analysis of the ICSS algorithm, some CEC 2017 benchmark functions have been used further to examine the performance and effectiveness of the developed algorithm. Therefore, in this subchapter, a comparison of the results obtained on the CEC 2017 functions between the ICSS algorithm and other well-known algorithms from the literature will be presented.

This small test will be executed considering a representative set of the CEC 2017 test functions, which are presented in detail in Table 5.14 in the appendix. This evaluation covers approximately 30 executions in 30 dimensions, where the stopping criterion is defined at a total of 1000 iterations, in the same way as it was done with the rest of the experiments in the benchmark functions. Additionally, the metaheuristic algorithms used in this comparison received a small change; that is, one algorithm was replaced by another known in the literature in order to provide a greater diversity of comparison between the different performance tests. The algorithms used in this comparison are the Gray Wolf Operator (GWO) [14], Differential Evolution (DE) [15], Particle Swarm Optimization (PSO) [16], Crow Search Algorithm (CSA) [17], Moth-Flame Optimization Algorithm (MFO) [18], and finally the Artificial Bee Colony Algorithm (ABC) [27]. The operating parameters with which the algorithms worked were the same as those recommended by their authors.

Table 5.7 shows the results obtained from this experiment, where the metrics used are the best fitness value (f_{BEST}), the mean fitness value $\left(\bar{f}\right)$ and the standard deviation $\left(\sigma_f\right)$. In addition, to facilitate the identification of the most efficient algorithms, the same procedure was followed as with the other experiment tables because there are functions where more than one algorithm had optimal results, achieving the best possible solution in the search space. The entries that show better performance are highlighted in bold. As can be seen, in most of the functions used in this representative

Table 5.7 Results of the minimization of the representative set of the CEC 2017 test functions in 30 dimensions

		GWO	DE	PSO	CSA	MFO	ABC	ICSS
f_1	f_{Best}	1.76E+09	2.31E+07	4.54E+09	5.73E+10	9.00E+09	1.16E+07	**8.41E+05**
	$\overline{f}$	1.69E+09	1.46E+06	3.93E+09	5.73E+10	9.29E+09	8.68E+06	1.62E+06
	σ_f	1.22E+09	8.21E+07	2.90E+09	8.04E+09	6.01E+09	1.25E+07	6.29E+05
f_2	f_{Best}	5.90E+02	5.14E+02	9.64E+02	1.51E+04	9.21E+02	**4.87E+02**	5.09E+02
	$\overline{f}$	5.60E+02	5.16E+02	8.29E+02	1.56E+04	8.45E+02	4.79E+02	5.59E+02
	σ_f	8.85E+01	1.25E+01	4.80E+02	3.02E+03	3.97E+02	4.43E+01	1.73E+01
f_3	f_{Best}	**6.05E+02**	6.56E+02	6.21E+02	9.37E+02	6.96E+02	7.36E+02	7.84E+02
	$\overline{f}$	6.04E+02	6.58E+02	6.13E+02	9.34E+02	7.00E+02	7.37E+02	8.41E+02
	σ_f	2.18E+01	1.27E+01	2.91E+01	3.43E+01	5.29E+01	1.47E+01	2.36E+01
f_4	f_{Best}	6.08E+02	**6.00E+02**	6.13E+02	6.90E+02	6.32E+02	6.03E+02	6.62E+02
	$\overline{f}$	6.08E+02	6.00E+02	6.12E+02	6.90E+02	6.32E+02	6.03E+02	6.80E+02
	σ_f	3.67E+00	2.83E-13	7.58E+00	6.14E+00	1.01E+01	7.70E-01	9.79E+00
f_5	f_{Best}	9.53E+05	1.44E+06	8.03E+05	2.04E+07	3.92E+06	1.33E+07	**4.98E+05**
	$\overline{f}$	6.89E+05	1.25E+06	4.86E+05	1.90E+07	8.41E+05	1.20E+07	2.27E+06
	σ_f	1.09E+06	7.26E+05	8.00E+05	1.15E+07	8.03E+06	6.03E+06	8.69E+05
f_6	f_{Best}	2.39E+03	2.45E+03	2.44E+03	2.73E+03	2.49E+03	2.54E+03	**2.28E+03**
	$\overline{f}$	2.38E+03	2.45E+03	2.43E+03	2.74E+03	2.50E+03	2.54E+03	2.50E+03
	σ_f	2.80E+01	1.32E+01	2.72E+01	3.32E+01	3.71E+01	1.29E+01	1.12E+02
f_7	f_{Best}	4.89E+03	8.64E+03	5.23E+03	9.02E+03	5.99E+03	1.03E+04	**2.33E+03**
	$\overline{f}$	5.37E+03	9.25E+03	5.80E+03	9.26E+03	6.62E+03	1.03E+04	2.44E+03
	σ_f	1.83E+03	2.01E+03	1.51E+03	8.37E+02	1.58E+03	2.74E+02	1.70E+03
f_8	f_{Best}	**2.75E+03**	2.80E+03	2.92E+03	3.39E+03	2.83E+03	2.90E+03	2.76E+03
	$\overline{f}$	2.75E+03	2.80E+03	2.91E+03	3.39E+03	2.82E+03	2.90E+03	3.01E+03
	σ_f	3.42E+01	1.94E+01	6.80E+01	6.40E+01	2.56E+01	1.21E+01	1.30E+02
f_9	f_{Best}	2.93E+03	3.00E+03	3.11E+03	3.58E+03	2.98E+03	3.07E+03	**2.61E+03**
	$\overline{f}$	2.91E+03	3.00E+03	3.09E+03	3.60E+03	2.98E+03	3.07E+03	3.14E+03
	σ_f	6.25E+01	1.10E+01	9.29E+01	8.71E+01	3.12E+01	1.32E+01	2.05E+02
f_{10}	f_{Best}	2.97E+03	**2.90E+03**	3.03E+03	6.08E+03	3.29E+03	2.91E+03	**2.90E+03**
	$\overline{f}$	2.96E+03	2.90E+03	2.97E+03	6.00E+03	3.16E+03	2.91E+03	2.94E+03
	σ_f	5.32E+01	1.21E+01	1.56E+02	7.89E+02	3.80E+02	8.61E+00	1.35E+01
f_{11}	f_{Best}	4.58E+03	5.04E+03	5.59E+03	1.07E+04	5.72E+03	5.65E+03	**2.84E+03**
	$\overline{f}$	4.62E+03	5.06E+03	5.65E+03	1.07E+04	5.73E+03	5.71E+03	2.98E+03
	σ_f	3.41E+02	2.06E+02	8.45E+02	7.52E+02	4.69E+02	1.64E+02	2.06E+02
f_{12}	f_{Best}	3.43E+03	3.35E+03	3.66E+03	6.69E+03	4.07E+03	3.30E+03	**3.27E+03**
	$\overline{f}$	3.40E+03	3.35E+03	3.51E+03	6.81E+03	3.79E+03	3.30E+03	3.31E+03
	σ_f	1.40E+02	5.70E+01	4.31E+02	1.01E+03	7.80E+02	3.16E-01	1.76E+01

set, the ICSS algorithm demonstrates superior performance compared to the rest of the algorithms used. This is possible thanks to its strategy balancing mechanism that ensures varied search patterns and promotes an adaptive and dynamic process, crucial characteristics to obtain optimal performance in these CEC 2017 functions due to the complexity of the search spaces of each of the functions that comprise it. Although the ICSS algorithm shows suboptimal performance in some of the test functions, the results obtained are constant or close to the best value of the search space. In other words, the ICSS algorithm is the second-best algorithm, which gives us a better overview of the capabilities of the algorithm, showing that it can produce good approximations to the best solution of optimization problems where different approaches are worked on in its search spaces.

Similarly to the benchmark functions presented in the previous subchapter, it is necessary to perform a statistical analysis of the results of the CEC 2017 test functions. As in the previous experimentation, the Wilcoxon signed-rank test was used to carry out this analysis. The results used to carry out this analysis are the same as those presented in Table 5.7, which are those corresponding to the CEC 2017 functions in 30 dimensions. The results obtained by the Wilcoxon test are detailed in Table 5.8. This table, like its predecessor in the previous subchapter, provides a pair-wise comparison between the ICSS algorithm and another metaheuristic algorithm, allowing for statistical validity of the experiments performed using the CEC 2017 test functions. As before, Table 5.8 employs the same symbols to represent the results of the Wilcoxon test, these being the symbols ▲, ▼, and ►. Likewise, the symbol ▲ represents that the ICSS algorithm outperformed the other competing algorithm, while the symbol ▼ represents that the other competing algorithm outperformed the ICSS algorithm. Finally, the symbol ► represents that the Wilcoxon test found no significant statistical differences between both algorithms.

According to the p-values shown in Table 5.8, the ICSS algorithm shows a performance superiority against the CSA algorithm since, in all the values obtained from all the functions studied, the ICSS technique outperforms its competitor. Furthermore, in functions f_6 and f_8, the ABC algorithm shows a performance similar to ICSS. At the same time, it can be observed that the GWO algorithm shows a performance similar to ICSS in function f_2. The Wilcoxon test also indicates that there is no significant difference between our algorithm and the PSO algorithm in functions f_6, f_8 and f_9. In the case of functions f_1 and f_6, the DE algorithm exhibits a performance similar to ICSS. In turn, the comparison of the ICSS and MFO algorithms does not show a significant difference in function f_6. The developed ICSS algorithm maintains its efficiency, robustness and scalability across most of the evaluated functions since in functions $f_3, f_4, f_5, f_7, f_{10}, f_{11}$ and f_{12}, the performance of the algorithm exceeds the performance of its competitors. Furthermore, only in three functions, these being functions f_1, f_2 and f_9, is there another metaheuristic algorithm that presents a similar performance, and in two functions, these being functions f_6 and f_8, there is more than one algorithm where the test does not detect a significant statistical difference between the obtained values.

In summary, the results of the Wilcoxon test shown in Table 5.8 authenticate the effectiveness of the ICSS algorithm in terms of accuracy and robustness, just as the

Table 5.8 Results of the Wilcoxon analysis of the CEC 2017 functions considering the values in Table 5.7

ICSS VS	GWO	DE	PSO	CSA	MFO	ABC
$f_1(x)$	3.01E-11▲	8.30E-01►	5.55E-10▲	3.01E-11▲	5.55E-10▲	3.49E-09▲
$f_2(x)$	2.74E-01►	9.14E-10▲	2.11E-07▲	3.01E-11▲	7.37E-08▲	2.94E-08▲
$f_3(x)$	3.00E-11▲	2.96E-11▲	3.00E-11▲	7.34E-11▲	1.14E-10▲	3.00E-11▲
$f_4(x)$	2.89E-11▲	1.20E-12▲	2.96E-11▲	7.20E-04▲	2.96E-11▲	1.99E-11▲
$f_5(x)$	3.44E-06▲	1.45E-04▲	1.60E-07▲	6.72E-10▲	3.77E-04▲	3.33E-11▲
$f_6(x)$	4.51E-04▲	1.29E-01►	5.32E-02►	2.91E-11▲	5.99E-01►	6.51E-01►
$f_7(x)$	9.01E-06▲	3.49E-08▲	6.61E-08▲	7.81E-10▲	3.30E-07▲	2.94E-11▲
$f_8(x)$	9.62E-10▲	3.49E-06▲	1.73E-01►	2.95E-11▲	2.65E-04▲	1.51E-01►
$f_9(x)$	3.90E-04▲	1.25E-03▲	7.96E-01►	2.93E-11▲	7.38E-04▲	7.90E-03▲
$f_{10}(x)$	1.89E-02▲	3.17E-10▲	2.15E-02▲	2.50E-11▲	3.66E-05▲	5.18E-10▲
$f_{11}(x)$	2.96E-11▲	2.96E-11▲	2.97E-11▲	2.93E-11▲	2.97E-11▲	2.96E-11▲
$f_{12}(x)$	4.50E-09▲	1.10E-03▲	2.78E-11▲	2.80E-11▲	2.38E-09▲	6.21E-05▲
Total ▲	12	11	10	13	12	11
Total ►	1	2	3	0	1	2
Total ▼	0	0	0	0	0	0

benchmark functions used in the other experiment section demonstrated. Although the GWO, DE, PSO, CSA and ABC algorithms show some similar results to the ICSS algorithm, the latter maintains its effectiveness, robustness and scalability in most of the evaluated scenarios, giving rise to its competitiveness in the literature.

In the same way as a convergence analysis was performed for the test functions of the performance analysis based on the data in Table 5.5, a convergence analysis was also performed for the CEC 2017 test functions. A brief representative set of the CEC 2017 convergence graphs is shown below in Fig. 5.8. These graphs were obtained from the CEC 2017 test function results in 30 dimensions, as shown in Table 5.7. Similar to the previous convergence analysis, looking at the data shown in Fig. 5.8, we can see that the ICSS algorithm again shows fast convergence, which is still attributed to its unique mechanism, in which each particle balances its individual I-LS strategy with the collective C-LS strategy guided by its respective counters.

To conclude, while the ICSS algorithm offers several advantages, it also faces limitations. Steering and coordinating two different strategies can lead to several opportunities but with a complicated implementation. This dual approach could also be difficult when performing parameter tuning. That is, the algorithm may require a larger number of iterations to reach an optimal solution. This greater demand, in terms of more iterations, is derived from the behavior of the particles guided by their counters since these same counters are the ones that, by determining the dynamics of each strategy, add a layer of complexity to the execution. Consequently, this complexity, although it can be beneficial to navigate optimization problem spaces

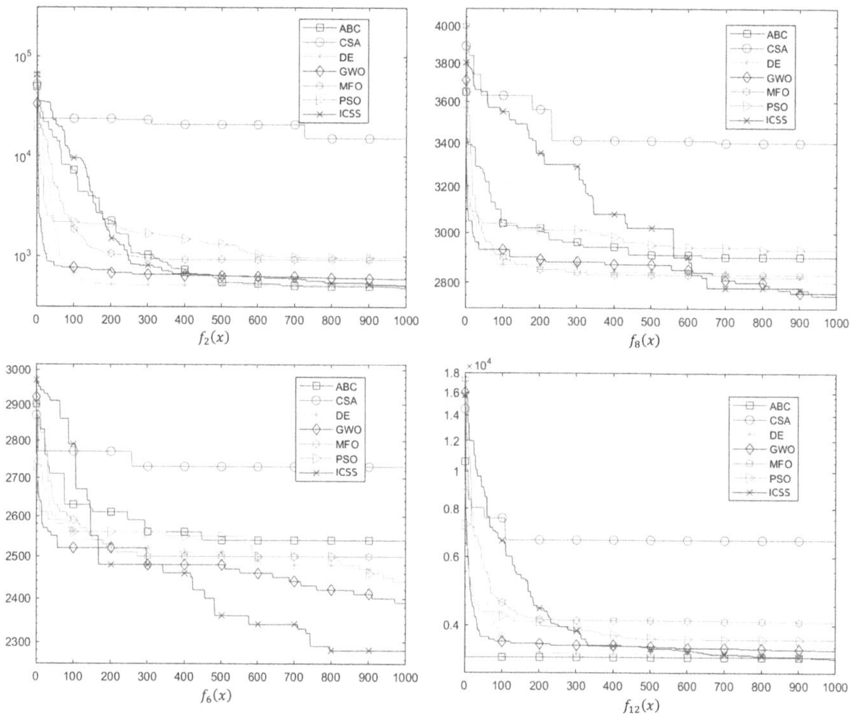

Fig. 5.8 Convergence graphics of the CEC 2017 functions f_2, f_6, f_8 and f_{12} in 30 dimensions

characterized by numerous local optima and high dimensionality, could translate into low efficiency in simpler spaces.

5.6.4 Engineering Problems

The main objective of an optimization algorithm is to find the best possible solution for a given problem. Optimization problems are paramount in various sectors, such as medicine, engineering, and economics, among others, where several fundamental challenges are framed in this context. To test the effectiveness of metaheuristic methods in practical scenarios, engineering design problems are often used, such as the three-bar truss design problem, the pressure vessel problem, and the welded beam problem. This subchapter is dedicated to the application of the ICS algorithm to a series of engineering design problems, thus evaluating its feasibility and effectiveness in real-world scenarios. The mathematical descriptions of each of these problems are detailed in Table 5.15 in the appendix. This analysis not only tests the applicability of the algorithm but also shows its potential to address more complex and specific problems in the industry.

In the analysis of the engineering design problems, the same metaheuristic algorithms as in the first performance analysis shown above were used. Each experiment was performed with a stopping criterion set at 1000 iterations. Also, given the stochastic nature of the metaheuristic algorithms, each engineering problem was run 30 times independently to ensure the reliability of the results. The analysis details the results of each of the problems, considering their decision variables, their restrictions and the specific fitness values. In addition, in this analysis, the statistical evaluation includes the worst fitness value, the mean, the standard deviation, and the best fitness value in order to provide a better overall understanding.

- *Design problem of a three-bar truss*
  This optimization problem focuses on a two-dimensional search space, with the specific objective of minimizing the structure of a three-bar loaded truss [28]. A drawing representing this truss structure is presented in Fig. 5.9. The optimization is subject to three constraints, which are detailed in Table 5.15 of the appendix. The results of this optimization problem are summarized in Table 5.9. These results demonstrate that the proposal is successful in solving this problem since it shows that the ICSS algorithm successfully finds the optimal solution. The ability of the method to effectively handle the constraints and complexity of this problem highlights its practical application and efficiency in solving tangible engineering problems.
- *Pressure Vessel Design Problem*
  The Pressure Vessel Design Problem [29], which is represented in Fig. 5.10, focuses on the optimization of a compressed air storage tank. The objective is to minimize the total cost of material, assembly and welding, considering several design parameters: vessel thickness, head thickness, inner radius and vessel length. A more detailed mathematical description of this problem and its constraints can be seen in Table 5.15 of the appendix. The results of applying the ICSS algorithm

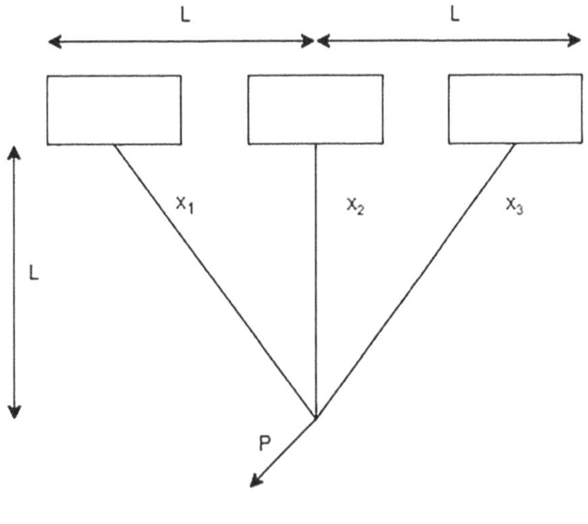

Fig. 5.9 Model of the design problem of a three-bar truss

Table 5.9 Results of the design problem of a three-bar truss

Algorithm	Worst	Media	Std	Best
CSA	1.7593e+02	1.7593e+02	4.9056e-09	**1.7593e+02**
DE	1.7593e+02	1.7593e+02	2.4984e-13	**1.7593e+02**
GA	1.7625e+02	1.7596e+02	7.1200e-02	**1.7593e+02**
GWO	1.7593e+02	1.7593e+02	1.5000e-03	**1.7593e+02**
MFO	1.7608e+02	1.7595e+02	3.9600e-02	**1.7593e+02**
PSO	1.7593e+02	1.7593e+02	1.0000e-04	**1.7593e+02**
ICSS	1.7655e+02	1.7608e+02	7.3500e-02	**1.7593e+02**

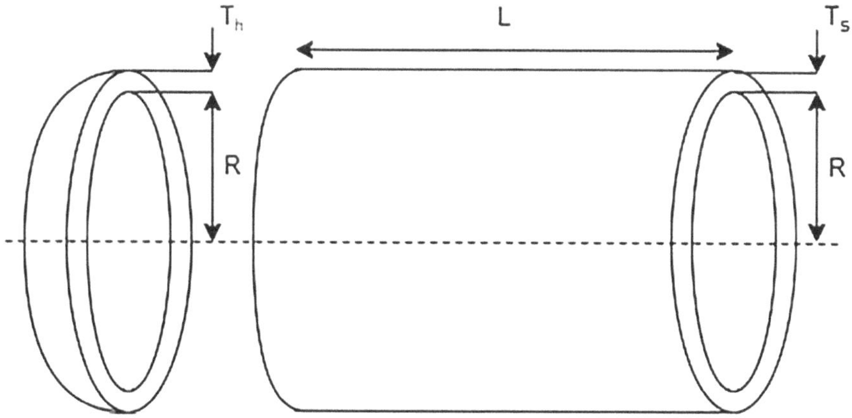

Fig. 5.10 Model of the pressure vessel design problem

to this problem are shown in Table 5.10. From the results, it can be seen that the ICSS algorithm. However, it does not always give the most optimal solution in the search space, but it still obtains suboptimal results and thus demonstrates greater efficiency than other algorithms.

Table 5.10 Pressure vessel design problem results

Algorithm	Worst	Media	Std	Best
CSA	5.9936e+03	5.9070e+03	2.0720e+01	5.8913e+03
DE	6.3496e+03	5.9615e+03	8.8387e+01	5.8962e+03
GA	7.5133e+03	6.6056e+03	2.9555e+02	6.3184e+03
GWO	7.2226e+03	5.9459e+03	2.4143e+02	5.8890e+03
MFO	7.3190e+03	6.3161e+03	4.6907e+02	**5.8853e+03**
PSO	7.3190e+03	6.2374e+03	5.3025e+02	**5.8853e+03**
ICSS	1.3754e+04	7.4929e+03	1.9734e+02	5.9630e+03

- *Welded beam design problem*
  The welded beam design problem [30], shown in Fig. 5.11, focuses on minimizing manufacturing costs, considering the optimization of four variables that are considered key in the design: width, length, depth and thickness. This problem is subject to seven constraints, resulting in this problem representing a complex scenario considered typical of engineering in the world. The detailed specifications of the welded beam problem are detailed in Table 5.15 in the appendix. Table 5.11 shows the details of the results of the different algorithms used to solve this challenge. The analysis of these data reveals that the ICSS method outperforms other metaheuristic methods, achieving the best solution for this specific problem. This result not only demonstrates the effectiveness of the proposal but also confirms its applicability and superiority in engineering problems, specifically in the field of design optimization.

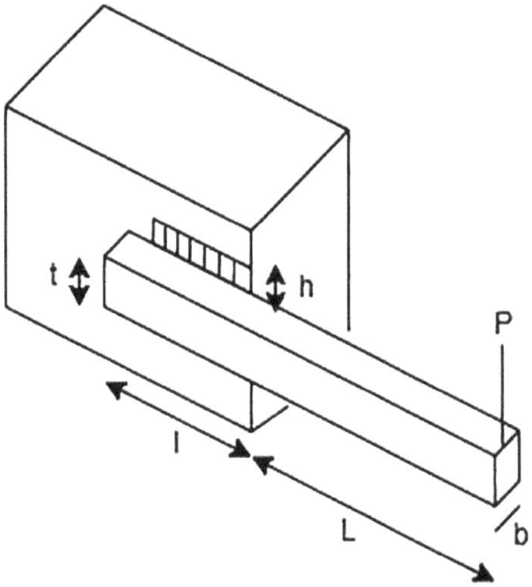

Fig. 5.11 Model of the Welded beam design problem

Table 5.11 Welded beam design problem results

Algorithm	Worst	Media	Std	Best
CSA	1.8285e+00	1.7569e+00	2.1200e-02	1.7333e+00
DE	1.9814e+00	1.8158e+00	6.2200e-02	1.7481e+00
GA	1.0000e+300	2.3333e+299	1.0000e+999	2.8176e+00
GWO	1.0000e+300	3.3333e+299	1.0000e+999	1.7252e+00
MFO	3.7761e+00	1.8830e+00	4.1390e-01	17249e+00
PSO	1.7261e+00	1.7249e+00	2.2814e-04	1.7249e+00
ICSS	2.7215e+00	2.0886e+00	2.5980e-01	**1.6509e+00**

5.6.5 NP-Hard Engineering Problems

In computer science, the acronym NPC or NP stands for Nondeterministic Polynomial Time (NPT) and refers to a complex class of decision problems [31]. The main characteristic of these problems is that checking the solution is much easier than finding the solution from scratch. Several combinatorial computational problems are considered NP-Hard problems, some examples of which are the Boolean Satisfiability Problem (SAT) [32], the Temporal Knapsack Problem (TKP) [33] and the Traveling Salesman Problem (TSP) [34]. The Traveling Salesman Problem (TSP) is used in the following to evaluate the proposal's applicability to various contemporary challenges by demonstrating its relevance and versatility.

The traveling salesman problem (TSP) is a combinatorial problem that focuses on the set of cities and the cost of travel between them. In other words, it is based on the idea that a salesman spends his time visiting N cities (nodes) that are connected to each other by several paths (edges), considering that each city must be visited only once. The trip must end where it started. The optimization problem focuses on finding the best possible path to travel through all the cities and returning to the starting point where the cost or distance of the trip is minimized [35]. The following Eq. 5.7 is the function that represents this optimization problem, and it is subject to four constraints (Eqs. 5.8–5.11) that are detailed below:

$$f(x) = \sum_{i=1}^{N} \sum_{\substack{i=1 \\ i \neq j}}^{N} c_{ij} x_{ij} \tag{5.7}$$

$$\sum_{\substack{i=1 \\ i \neq j}}^{N} x_{ij} = 1, \forall j \in \{1, \ldots, N\} \tag{5.8}$$

$$\sum_{\substack{j=1 \\ i \neq j}}^{N} x_{ij} = 1, \forall i \in \{1, \ldots, N\} \tag{5.9}$$

$$\sum_{\substack{i \in s \\ j \in s \\ i \neq j}}^{N} x_{ij} \geq 1, \forall s \in V \tag{5.10}$$

$$x_{ij} = 0 \, or \, 1 (i, j) \in A \tag{5.11}$$

Table 5.12 Results of the TSP problem

Algorithm	Worst	Media	Std	Best	Best route
CSA	661	596.63	55.34	494	QHSDEBOALIJKCMFGTPNR
DE	629	565.63	35.87	471	TNOPSIKLGAJDEMRQBHCF
GA	637	572.9	33.98	478	TSNKRFDEBPMLGAIHOLQC
GWO	629	563.53	48.43	434	QNSFTLIAGEPMBOHKRDCJ
MFO	654	592.26	42.14	476	CKNBOPTJMQHSIEGRDALF
PSO	636	564.23	44.41	453	PRKJNFTMIDSQCGALOEHB
EBIC	630	570.06	47.84	**406**	TIEOMBPDAQGLFRJKNHSC

where x_{ij} represents the existence of a path between the cities i and j, where is equal to 1 if there is a path to go from city i to city j, and where is equal to 0 otherwise. Then c_{ij} is the Euclidean distance from node i to node j. Additionally, $V = \{v_1, v_2, \ldots, v_n\}$ is the set of cities, and $A = \{(r, s) : r, s \in V\}$ is the set of edges connecting those cities. The example of the TSP problem used to analyze the proposed algorithm can be defined by the adjacent matrix shown in Table 5.16 in the Appendix. This case is formulated by a matrix of twenty cities as nodes with different costs as vertices connecting these same nodes. For the analysis, continuous solutions are mapped into permutations of cities, where each solution represents a combination of interconnected nodes. In addition, each agent selects an initial city as a starting node q_0 and from there, the operators of the I-LS and C-LS strategies are applied to minimize the sum of the costs, i.e., the optimization function. It is important to note that each agent represents a feasible combination of cities and connections (nodes and edges), so this process is repeated for each agent.

The analysis of the results is carried out by comparing the results of the same algorithms used in the previous engineering design problems, considering 30 independent executions. The results of the TSP problem are shown in Table 5.12. The best and worst results, as well as their mean and standard deviation, can be observed in the table. The best route is also presented. These results demonstrate that the proposal successfully finds the optimal solution to the problem, providing the best route with the lowest cost. The effectiveness of the algorithm in handling the constraints and complexities of this problem exhibits its practical application and its efficiency in terms of exploration and exploitation. Finally, it is worth mentioning that the ICSS algorithm can obtain high-quality solutions based on its operators that allow for generating the balance between individual and collective strategies.

5.7 Conclusion

In this chapter, a new metaheuristic algorithm named "Individual and Collective Search Strategy" (ICSS) was studied to solve optimization problems. The algorithm is based on the combination of individual and collective strategies for the particles in the search space. Each particle uses two strategies: an independent search and a collective search based on the best solution found so far. With all this, the algorithm considers the search space as a learning problem, where the particles are required to find a balance between the strategies. This is achieved with the support of a counter, which is assigned to each particle with a randomly generated limit, and thanks to this counter, each particle establishes a frequency of use of the strategies. This mechanism gives rise to diverse search patterns, along with a more dynamic and acceptable process.

The ICSS algorithm shows a great performance associated with its two strategies: through the individual strategy, the particle exploits the information collected by itself; in this way, the search area assigned to it is exploited to search for the optimal solution; and through the collective strategy the particle explores the search space thanks to the information provided by the rest of the population, especially the particle that has the best solution so far, and with this, it can determine promising areas and where not in order to better explore the search space. With this balance, the particle uses its own information. It searches its area for better solutions, and from time to time, it will look at the rest of the particles to obtain more information about the search space and thus improve the quality of the solutions by moving to more promising areas and avoiding undesirable areas.

The effectiveness of the proposed algorithm was evaluated alongside a series of metaheuristic algorithms, among which some stand out, such as DE, GA, GWO, PSO, etc. This comparative analysis was carried out with 20 functional benchmarks, among which there are unimodal, multimodal and hybrid functions and four engineering tests. The results of these experiments indicate that the proposal outperforms these metaheuristic algorithms, demonstrating a higher solution quality and a faster convergence rate, as well as the effectiveness of the algorithm in handling the restrictions and complexities that engineering problems exhibit, thereby demonstrating its practical application. All of this highlights the potential that the algorithm has as a more efficient and effective approach to solving complex optimization problems.

Appendix

See Tables 5.13, 5.14, 5.15 and 5.16.

Table 5.13 Full description of benchmark functions

Nombre	Función	S	Mínimo
Ackley	$f_1(\boldsymbol{x}) = -20\, exp\left(-0.2\sqrt{\frac{1}{n}\sum_{i=1}^{n} x_i^2}\right) - exp\left(\frac{1}{n}\sum_{i=1}^{n} cos(2\pi x_i)\right) + 20 + e$	$[-32, 32]$	$\boldsymbol{x}^* = (0, \ldots, 0)$ $f(\boldsymbol{x}^*) = 0$
Dixon-price	$f_2(x) = (x_i - 1)^2 + \sum_{i=2}^{d} i(2x_i^2 - x_{i-1})^2$	$[-10, 10]$	$x_i = 2^{-\frac{2^i-2}{2^i}}$ $f(\boldsymbol{x}^*) = 0$
Griewank	$f_3(x) = \sum_{i=1}^{d} \frac{x_i^2}{4000} - \prod_{i=1}^{d} \cos\left(\frac{x_i}{\sqrt{i}}\right) + 1$	$[-600, 600]$	$\boldsymbol{x}^* = (0, \ldots, 0)$ $f(\boldsymbol{x}^*) = 0$
Infinity	$f_4(x) = \sum_{i=1}^{n} x_i^6 * (\sin(x_i) + 2)$	$[-1, 1]$	$\boldsymbol{x}^* = (0, \ldots, 0)$ $f(\boldsymbol{x}^*) = 0$
Levy	$f_5(x) = sin^2(\pi w_i) + \sum_{i=1}^{d-1} (w_i - 1)^2[1 + 10sin^2(\pi w_i + 1)] + (w_d - 1)^2[1 + sin^2(\pi w_d)]$ $w_i = 1 + \frac{x_i - 1}{4}$	$[-10, 10]$	$\boldsymbol{x}^* = (1, \ldots, 1)$ $f(\boldsymbol{x}^*) = 0$
Mishra 11	$f_6(x) = \left[\frac{1}{n}\sum_{i=1}^{n} \lvert x_i\rvert - \left(\prod_{i=1}^{n} \lvert x_i\rvert\right)^{\frac{1}{n}}\right]^2$	$[-10, 10]$	$\boldsymbol{x}^* = (0, \ldots, 0)$ $f(\boldsymbol{x}^*) = 0$
Multimodal	$f_7(x) = \sum_{i=1}^{n} \lvert x_i\rvert * \prod_{i=1}^{n} \lvert x_i\rvert$	$[-10, 10]$	$\boldsymbol{x}^* = (0, \ldots, 0)$ $f(\boldsymbol{x}^*) = 0$
Penalized 02	$f_8(x) = 0.1\left\{sin^2(3\pi x_1) + \sum_{i=1}^{n-1} (x_i - 1)^2\left[1 + sin^2(3\pi x_{i+1})\right] + (x_n - 1)^2\left[1 + sin^2(2\pi x_n)\right]\right\} + \sum_{i=1}^{n} u(x_i, 5{,}100{,}4)$ $u(x_i, a, k, m) = \begin{cases} k(x_i - a)^m \; if \; x_i > a \\ 0 \; if \; -a \le x_i \le a \\ k(-x_i - a)^m \; if \; x_i < -a \end{cases}$	$[-50, 50]$	$\boldsymbol{x}^* = (1, \ldots, 1)$; $f(\boldsymbol{x}^*) = 0$
Plateau	$f_9(x) = 30 + \sum_{i=1}^{n} \lfloor x_i \rfloor$	$[-5.12, 5.12]$	$\boldsymbol{x}^* = (0, \ldots, 0)$ $f(\boldsymbol{x}^*) = 30$

(continued)

Table 5.13 (continued)

Nombre	Función	S	Mínimo
Powell	$f_{10}(x) = \sum_{i=1}^{d/4}\left[(x_{4i-3} + 10x_{4i-2})^2 + 5(x_{4i-1} - x_{4i})^2 + (x_{4i-2} - 2x_{4i-1})^4 + 10(x_{4i-3} - x_{4i})^4\right]$	[−4, 5]	$\boldsymbol{x}^* = (0, \ldots, 0)$ $f(\boldsymbol{x}^*) = 0$
Qing	$f_{11}(x) = \sum_{i=1}^{n} (x_i^2 - i)^2$	[−500, 500]	$\boldsymbol{x}^* = \left(\pm\sqrt{i}, \ldots, \pm\sqrt{i}\right)$ $f(\boldsymbol{x}^*) = 0$
Quartic	$f_{12}(\boldsymbol{x}) = \sum_{i=1}^{n} ix_i^4 + random(0,1)$	[−1.28, 1.28]	$\boldsymbol{x}^* = (0, \ldots, 0)$ $f(\boldsymbol{x}^*) = 0$
Rastrigin	$f_{13}(x) = 10d + \sum_{i=1}^{d}\left[x_i^2 - 10\cos(2\pi x_i)\right]$	[−5.12, 5.12]	$\boldsymbol{x}^* = (0, \ldots, 0)$ $f(\boldsymbol{x}^*) = 0$
Rosenbroc	$f_{14}(x) = \sum_{i=1}^{n-1}\left[100(x_i^2 - x_{i+1})^2 + (x_i - 1)^2\right]$	[−5, 10]	$\boldsymbol{x}^* = (1, \ldots, 1)$ $f(\boldsymbol{x}^*) = 0$
Schewel 22	$f_{15}(x) = \sum_{i=1}^{n} \lvert x_i \rvert + \prod_{i=1}^{n} \lvert x_i \rvert$	[−100, 100]	$\boldsymbol{x}^* = (0, \ldots, 0)$ $f(\boldsymbol{x}^*) = 0$
Rotated hyper ellipsoid (Rothyp)	$f_{16}(x) = \sum_{i=1}^{n} \sum_{j=1}^{i} x_j^2$	[−65.536, 65.536]	$\boldsymbol{x}^* = (0, \ldots, 0)$ $f(\boldsymbol{x}^*) = 0$
Sphere	$f_{17}(x) = \sum_{i=1}^{n} x_i^2$	[−1, 1]	$\boldsymbol{x}^* = (0, \ldots, 0)$ $f(\boldsymbol{x}^*) = 0$
SumPow	$f_{18}(x) = \sum_{i=1}^{n} \lvert x_i \rvert^{i+1}$	[−1, 1]	$\boldsymbol{x}^* = (0, \ldots, 0)$ $f(\boldsymbol{x}^*) = 0$
Hybrid 01	$f_{19}(x) = f_{17}(x) + f_{13}(x) + f_{16}(x)$	[−100, 100]	$\boldsymbol{x}^* = (0, \ldots, 0)$ $f(\boldsymbol{x}^*) = 0$
Hybrid 02	$f_{20}(x) = f_{13}(x) + f_3(x) + f_{14}(x)$	[−100, 100]	$\boldsymbol{x}^* = (0, \ldots, 0)$ $f(\boldsymbol{x}^*) = n - 1$

Table 5.14 Description of the CEC 2017 test functions used

Nombre	Función	S	Mínimo
Shifted and rotated bent cigar function	$f_1(x) = x_1^2 + 10^6 \sum_{i=2}^{D} x_i^2$ $F_1(x) = f_1(M(x - o_1)) + F_1{}^*$	$[-100, 100]^D$	$f(\boldsymbol{x}^*) = 100$
Shifted and rotated Rosenbrock function	$f_2(x) = \sum_{i=1}^{D-1}\left(100\left(x_i^2 - x_{i+1}\right)^2 + (x_i - 1)^2\right)$ $F_2(x) = f_2\left(M\left(\frac{2.048(x-o_2)}{100}\right) + 1\right) + F_2{}^*$	$[-100, 100]^D$	$f(\boldsymbol{x}^*) = 400$
Shifted and rotated Rastrigin function	$f_3(x) = \sum_{i=1}^{D}\left(x_i^2 - 10\,cos(2\pi x_i) + 10\right)$ $F_3(x) = f_3(M(x - o_3)) + F_3{}^*$	$[-100, 100]^D$	$f(\boldsymbol{x}^*) = 500$
Shifted and rotated expanded Schaffer F6 function	$g(x, y) = 0.5 + \frac{\left(sin^2\left(\sqrt{x^2+y^2}\right)-0.5\right)}{\left(1+0.001\left(x^2+y^2\right)\right)^2}$ $f_4(x) = g(x_1, x_2) + g(x_2, x_3) + \cdots + g(x_{D-1}, x_D) + g(x_D, x_1)$ $F_4(x) = f_4\left(M\left(\frac{0.5(x-o_4)}{100}\right)\right) + F_4{}^*$	$[-100, 100]^D$	$f(\boldsymbol{x}^*) = 600$
Hybrid function 6 (N = 5)	$F_5(x) = g_1(M_1Z_1) + g_2(M_2Z_2) + \cdots + g_N(M_NZ_N) + F_5{}^*(x)$ g_1: *High Conditioned Elliptic Function* g_2: *Ackley Function* g_3: *Rastrigin Function* g_4: *HGBat Function* g_5: *Discus Function*	$[-100, 100]^D$	$f(\boldsymbol{x}^*) = 1800$
Composition function 1 (N = 3)	$F_6(x) = \sum_{i=1}^{N}\left\{w_i * \left[\lambda_i g_i(x) + bias_i\right]\right\} + F_6{}^*$ $\lambda = \left[1, 1^{-6}, 1\right], bias = [0, 100, 200]$ g_1: *Rosenbrock Function* g_2: *High Conditioned Elliptic Function* g_3: *Rastrigin Function*	$[-100, 100]^D$	$f(\boldsymbol{x}^*) = 2100$
Composition function 2 (N = 3)	$F_7(x) = \sum_{i=1}^{N}\left\{w_i * \left[\lambda_i g_i(x) + bias_i\right]\right\} + F_7{}^*$ $\lambda = [1, 10, 1], bias = [0, 100, 200]$ g_1: *Rastrigin Function* g_2: *Griewank Function* g_3: *Modified Schewel Function*	$[-100, 100]^D$	$f(\boldsymbol{x}^*) = 2200$
Composition function 3 (N = 4)	$F_8(x) = \sum_{i=1}^{N}\left\{w_i * \left[\lambda_i g_i(x) + bias_i\right]\right\} + F_8{}^*$ $\lambda = [1, 10, 1, 1], bias = [0, 100, 200, 300]$ g_1: *Rosenbrock Function* g_2: *Ackley Function* g_3: *Modified Schewel Function* g_4: *Rastrigin Function*	$[-100, 100]^D$	$f(\boldsymbol{x}^*) = 2300$

(continued)

Table 5.14 (continued)

Nombre	Función	S	Mínimo
Composition function 4 (N = 4)	$F_9(x) = \sum_{i=1}^{N}\{w_i * [\lambda_i g_i(x) + bias_i]\} + F_9{}^*$ $\lambda = [10, 1^{-6}, 10, 1]$ $bias = [0, 100, 200, 300]$ g_1: *Ackley Function* g_2: *High Conditioned Elliptic Function* g_3: *Griewank Function* g_4: *Rastrigin Function*	$[-100, 100]^D$	$f(\boldsymbol{x}^*) =$ 2400
Composition function 5 (N = 5)	$F_{10}(x) = \sum_{i=1}^{N}\{w_i * [\lambda_i g_i(x) + bias_i]\} + F_{10}{}^*$ $\lambda = [10, 1, 10, 1^{-6}, 1]$ $bias = [0, 100, 200, 300, 400]$ g_1: *Rastrigin Function* g_2: *Happycat* g_3: *Ackley Function* g_4: *Discus Function* g_5: *Rosenbrock Function*	$[-100, 100]^D$	$f(\boldsymbol{x}^*) =$ 2500
Composition function 6 (N = 5)	$F_{11}(x) = \sum_{i=1}^{N}\{w_i * [\lambda_i g_i(x) + bias_i]\} + F_{11}{}^*$ $\lambda = [1^{-26}, 10, 1^{-6}, 10, 5^{-4}]$ $bias = [0, 100, 200, 300, 400]$ g_1: Expanded Schaffer g_2: *Modified Schewel Function* g_3: *Griewank Function* g_4: *Rosenbrock Function* g_5: *Rastrigin Function*	$[-100, 100]^D$	$f(\boldsymbol{x}^*) =$ 2600
Composition function 8 (N = 6)	$F_{12}(x) = \sum_{i=1}^{N}\{w_i * [\lambda_i g_i(x) + bias_i]\} + F_{12}{}^*$ $\lambda = [10, 10, 1^{-6}, 1, 5^{-4}]$ $bias = [0, 100, 200, 300, 400, 500]$ g_1: *Ackley Function* g_2: *Griewank Function* g_3: *Discus Function* g_4: *Rosenbrock Function* g_5: *Happycat* g_6: Expanded Schaffer	$[-100, 100]^D$	$f(\boldsymbol{x}^*) =$ 2800

Table 5.15 Full description of engineering problems

Nombre	Función	Rango(s)	Dim.
Three-bar truss design	$f_1(x) = \left(2 * \sqrt{2} * x_1 + x_2\right) * li$ $g(1) = \left(\frac{\sqrt{2}*x_1+x_2}{\sqrt{2}*x_1^2+2*x_1*x_2}\right) * p - \varepsilon$ $g(2) = \left(\frac{x_2}{\sqrt{x}*x_1^2+2*x_1*x_2}\right) * p - \varepsilon$ $g(3) = \left(\frac{1}{\sqrt{2}*x_1+x_2}\right) * p - \varepsilon$ $p = 2$ $\varepsilon = 2$ $li = 100$	$0 \le x_1 \le 1$ $0 \le x_2 \le 1$	n = 2
Pressure vessel design	$f_2(x) = 0.6224 * x_1 * x_3 * x_4 +$ $1.7781*x_2*x_3^2+3.1661*x_1^2*x_4+19.84*x_1^2*x_3$ $g(1) = -x_1 + 0.0193 * x_3$ $g(2) = -x_2 + 0.00954 * x_3$ $g(3) =$ $(-\pi * x_3^2 * x_4) - (\frac{4}{3} * \pi * x_3^3) + 1.296e06$ $g(4) = x_4 - 240$	$0.0625 \le x_1 \le 6.1875$ $0.0625 \le x_2 \le 6.1875$ $10 \le x_3 \le 200$ $10 \le x_4 \le 200$	n = 4
Welded-beam design	$f_3(x) =$ $1.1047*x_1^2*x_2+0.04811*x_3*x_4*(14+x_2)$ $g(1) = tF - tMax$ $g(2) = sigma\,F - sigma\,Max$ $g(3) = x_1 - x_4$ $g(4) =$ $0.10471*x_1^2+0.04811*x_3*x_4*(14+x_2)-5$ $g(5) = 0.125 - x_1$ $g(6) = delta\,F - delta\,Max$ $g(7) = P - PcF$ $M = P * (LL + \frac{x_2}{2})$ $R = \sqrt{\frac{x_2^2}{4} + (\frac{x_1+x_3}{2})^2}$ $J = 2 * \left(\sqrt{2} * x_1 * x_2 * \left(\frac{x_2^2}{12} + (\frac{x_1+x_3}{2})^2\right)\right)$ $sigmaF = \frac{6*P*LL}{x_3^2*x_4}$ $deltaf = \frac{4*P*LL^3}{E*x_3^3*x_4}$ $Aux = \left(\frac{4.013*E*\sqrt{\frac{x_3^2*x_4^6}{36}}}{LL^2}\right)$	$0.1 \le x_1 \le 2$ $0.1 \le x_2 \le 10$ $0.1 \le x_3 \le 10$ $0.1 \le x_4 \le 2$	n = 4

(continued)

Table 5.15 (continued)

Nombre	Función	Rango(s)	Dim.
	$PcF = Aux * \left(\frac{1-x_3}{2*LL}\right) * \left(\sqrt{\frac{E}{4*G}}\right)$ $Tprima = \frac{P}{\sqrt{2}*x_1*x_2}$ $Tbiprima = \frac{M*R}{J}$ $tF =$ $\sqrt{Tprima^2 + \frac{2*Tprima*Tbiprima*x_2}{2*R} + Tbiprima^2}$ $P = 6000$ $LL = 14$ $E = 30e6$ $G = 12e6$ $tMax = 13600$ $sigmaMax = 30000$ $deltaMax = 0.25$		

Table 5.16 Adjacency matrix of the TSP problem

	A	B	C	D	E	F	G	H	I	J	K	L	M	N	O	P	Q	R	S	T
A	0	82	91	13	92	64	10	28	55	96	97	16	98	96	49	81	15	43	92	80
B	82	0	96	66	4	85	94	68	76	75	40	66	18	71	4	28	5	10	83	70
C	91	96	0	32	96	4	44	39	77	80	19	49	45	65	71	76	28	68	66	17
D	13	66	32	0	12	50	96	35	59	23	76	26	51	70	90	96	55	14	15	26
E	92	4	96	12	0	85	26	82	25	93	35	20	26	62	48	36	84	59	55	92
F	64	85	4	50	85	0	29	76	76	39	57	8	6	54	78	94	13	57	47	2
G	10	94	44	96	26	29	0	34	17	80	32	53	17	61	27	66	69	75	46	9
H	28	68	39	35	82	76	34	0	23	92	16	83	54	100	8	45	11	97	1	78
I	55	76	77	59	25	76	17	23	0	82	87	9	40	26	81	44	92	19	27	15
J	96	75	80	23	93	39	80	92	82	0	14	87	58	55	15	86	63	36	52	41
K	97	40	19	76	35	57	32	16	87	14	0	8	24	13	19	24	42	5	91	95
L	16	66	49	26	20	8	53	83	9	87	8	0	50	49	34	91	37	12	79	39
M	98	18	45	51	26	6	17	54	40	58	24	50	0	25	41	10	14	95	96	58
N	96	71	65	70	62	54	61	100	26	55	13	49	25	0	6	24	36	83	2	5
O	49	4	71	90	48	78	27	8	81	15	19	34	41	6	0	17	65	74	65	46
P	81	28	76	96	36	94	66	45	44	86	24	91	10	24	17	0	55	30	75	19
Q	15	5	28	55	84	13	69	11	92	63	42	37	14	36	65	55	0	69	19	37
R	43	10	68	14	59	57	75	97	19	36	5	12	95	83	74	30	69	0	63	79
S	92	83	66	15	55	47	46	1	27	52	91	79	96	2	65	75	19	63	0	9
T	80	70	17	26	92	2	9	78	15	41	95	39	58	5	46	19	37	79	9	0

References

1. K. Zhang, Z. Yang, T. Başar, Multi-agent reinforcement learning: a selective overview of theories and algorithms, in *Handbook of Reinforcement Learning and Control*, eds. by K.G. Vamvoudakis, Y. Wan, F.L. Lewis, D. Cansever (Springer International Publishing, Cham, 2021), pp. 321–384. https://doi.org/10.1007/978-3-030-60990-0_12
2. K. Sörensen, F. Glover, Metaheuristics. Encycl. Oper. Res. Manag. Sci. **62**, 960–970 (2013)
3. F. Wang, H. Zhang, K. Li, Z. Lin, J. Yang, X.-L. Shen, A hybrid particle swarm optimization algorithm using adaptive learning strategy. Inf. Sci. (N. Y.) **436**, 162–177 (2018)
4. M. Van Harmelen, Personal learning environments, in *Sixth International Conference on Advanced Learning Technologies* (IEEE Computer Society, 2006), pp. 815–816
5. D. F. Dansereau, "7 - COOPERATIVE LEARNING STRATEGIES," in *Learning and Study Strategies*, C. E. Weinstein, E. T. Goetz, and P. A. Alexander, Eds., San Diego: Academic Press, 1988, pp. 103–120. https://doi.org/10.1016/B978-0-12-742460-6.50013-X.
6. R. Alterman, K. Harsch, Collaborative and individual learning, in *Proceedings of the 7th International Conference on Computer Supported Education* (2015), pp. 411–417
7. B. Chopard, M. Tomassini, *An Introduction to Metaheuristics for Optimization* (Springer, 2018)
8. B. Morales-Castañeda, D. Zaldivar, E. Cuevas, F. Fausto, A. Rodríguez, A better balance in metaheuristic algorithms: does it exist? Swarm Evol. Comput. **54**, 100671 (2020)
9. K. Marti, *Stochastic Optimization Methods*, vol. 2 (Springer, 2008)
10. K. Hussain, M.N. Mohd Salleh, S. Cheng, Y. Shi, Metaheuristic research: a comprehensive survey. Artif. Intell. Rev. **52**, 2191–2233 (2019)
11. L. Velasco, H. Guerrero, A. Hospitaler, A literature review and critical analysis of metaheuristics recently developed. Arch. Comput. Methods Eng. **31**(1), 125–146 (2024)
12. A. Obuchowicz, P. Pretki, sssIsotropic symmetric/spl alpha/-stable mutations for evolutionary algorithms, in *2005 IEEE Congress on Evolutionary Computation*, IEEE (2005), pp. 404–410
13. M. Jamil, X.-S. Yang, A literature survey of benchmark functions for global optimisation problems. Int. J. Math. Model. Numer. Optim. **4**(2), 150–194 (2013)
14. S. Mirjalili, S.M. Mirjalili, A. Lewis, Grey wolf optimizer. Adv. Eng. Softw. **69**, 46–61 (2014)
15. R. Storn, K. Price, *Differential Evolution-A Simple and Efficient Heuristic for Global Optimization over Continuous Spaces* (Kluwer Academic Publishers, 1997
16. J. Kennedy, R. Eberhart, Particle swarm optimization, in *Proceedings of ICNN'95-International Conference on Neural Networks*, IEEE (1995), pp. 1942–1948
17. A. Askarzadeh, A novel metaheuristic method for solving constrained engineering optimization problems: crow search algorithm. Comput. Struct. **169**, 1–12 (2016). https://doi.org/10.1016/j.compstruc.2016.03.001
18. S. Mirjalili, Moth-flame optimization algorithm: a novel nature-inspired heuristic paradigm. Knowl. Based Syst. **89**, 228–249 (2015)
19. J.H. Holland, Genetic algorithms and adaptation, in *Adaptive Control of Ill-Defined Systems* (1984), pp. 317–333
20. J. Holland, Genetic algorithms. Sci. Am. **267**(1), 66–73 (1992)
21. R.F. Woolson, Wilcoxon signed-rank test, in *Encyclopedia of Biostatistics*, vol. 8 (2005)
22. F. Wilcoxon, Individual comparisons by ranking methods, in *Breakthroughs in Statistics: Methodology and Distribution* (Springer, 1992), pp. 196–202
23. I.C.A. Oyeka, G.U. Ebuh, Modified Wilcoxon signed-rank test. Open J. Stat. **2**(2), 172–176 (2012)
24. W.J. Gutjahr, Convergence analysis of metaheuristics, in *Matheuristics: Hybridizing Metaheuristics and Mathematical Programming* (Springer, 2009), pp. 159–187
25. G. Wu, R. Mallipeddi, P.N. Suganthan, *Problem definitions and evaluation criteria for the CEC 2017 competition on constrained real-parameter optimization*. National University of Defense Technology, Changsha, Hunan, PR China and Kyungpook National University, Daegu, South Korea and Nanyang Technological University, Singapore, Technical Report (2017)

26. N.H. Awad, M.Z. Ali, J.J. Liang, B.Y. Qu, P.N. Suganthan, *Problem definitions and evaluation criteria for the CEC 2017 special session and competition on single objective bound constrained real-parameter numerical optimization*. Technical report, Nanyang Technological University Singapore Singapore (2016), pp. 1–34
27. D. Karaboga, B. Basturk, A powerful and efficient algorithm for numerical function optimization: artificial bee colony (ABC) algorithm. J. Glob. Optim. **39**, 459–471 (2007)
28. D. Tao, X. Wei, H. Huang, Application of improved fruit fly optimization algorithm in three bar truss, in *International Conference on Intelligent Computing* (Springer, 2022), pp. 785–801
29. S. Hassan, K. Kumar, C.D. Raj, K. Sridhar, Design and optimisation of pressure vessel using metaheuristic approach. Appl. Mech. Mater. **465**, 401–406 (2014)
30. K. Deb, Optimal design of a welded beam via genetic algorithms. AIAA J. **29**(11), 2013–2015 (1991)
31. N. Maaroju, Choosing the best heuristic for a np-problem (2009)
32. M.L. Littman, S.M. Majercik, T. Pitassi, Stochastic Boolean satisfiability. J. Autom. Reason. **27**, 251–296 (2001)
33. M. Bartlett, A.M. Frisch, Y. Hamadi, I. Miguel, S.A. Tarim, C. Unsworth, The temporal knapsack problem and its solution, in *Integration of AI and OR Techniques in Constraint Programming for Combinatorial Optimization Problems: Second International Conference, CPAIOR 2005*, Prague, Czech Republic, 31 May–1 June 2005, Proceedings 2 (Springer, 2005), pp. 34–48
34. K.L. Hoffman, M. Padberg, G. Rinaldi, Traveling salesman problem. Encycl. Oper. Res. Manag. Sci. **1**, 1573–1578 (2013)
35. R. Matai, S.P. Singh, M.L. Mittal, Traveling salesman problem: an overview of applications, formulations, and solution approaches. Travel. Salesm. Probl. Theory Appl. **1**(1), 1–25 (2010)

Chapter 6
Diversity-Opposition Hybridization of the Cheetah Optimizer for Global Optimization

6.1 Introduction

In the field of global optimization, metaheuristic algorithms have become a reliable alternative for solving diverse optimization problems for different fields of application [1]. Among diverse metaheuristic algorithms, "nature-inspired" computational algorithms have brought the development of novel, simple and powerful computational schemes [2–4] extracting their inspiration in observations of diverse living organisms, collective groups of animals, physical and chemical transformations of the matter and biological processes modeled as mathematical and computational operators to reach a goal [5]. The approach of natural processes in the computer sciences has inspired several researchers to generate new evolutionary strategies inspired by diverse behaviors found in diverse species to reach a common goal or the hunting cycles or prey and predators, approximated to diverse mathematical operators to generate diverse evolutionary operators.

Several metaheuristic models found in the literature are strongly influenced by different biological and natural behaviors, collective swarm strategies or even the principle of reproduction of the species for modeling their operators. Some interesting examples found in the literature include the Differential Evolution (DE) proposed by Storn and Price [6, 7], the Genetic Algorithm (GA) [7] and the Evolutionary Strategies (ES) proposed by Beyel and Schwefel [8] which follow the Darwin theory of evolution and reproduction of the species for the generation of new solutions across a space. Other interesting examples are the Particle Swarm Optimization (PSO) proposed by Kennedy and Eberhart [9], the Whale Optimization Algorithm (WOA) proposed by Mirjalili and Lewis [10], the Grey Wolf Optimizer (WOA) proposed by Mirjalili [11], the Artificial Bee Colony (ABC) algorithm proposed by Karaboga [12], the Bat Algorithm (BA) proposed by Yang [13], the Cuckoo Search (CS) proposed by Yang [14], the Crow Search Algorithm (CSA) proposed by Agushaka et al. [15] are inspired and influenced by collective swarms and groups of animals and living beings to reach a common goal. Other interesting bio-inspired

E. Cuevas et al., *Advanced Metaheuristics: Novel Approaches for Complex Problem Solving*, Studies in Computational Intelligence 1210,
https://doi.org/10.1007/978-3-031-89284-4_6

algorithms are those that are inspired by various physical and chemical phenomena by which matter is transformed in different states, giving as examples the Simulated Annealing (SA) methodology proposed by Kirkpatrick et al. [16], the Gravitational Search Algorithm (GSA) proposed by Rashedi et al. [17], the Harmony Search (HS) algorithm proposed by Geem et al. [18], the States of Matter search (SMS) algorithm proposed by Cuevas et al. [19] or the Kepler Optimization Algorithm (KOA) proposed by Abdel Basset et al. [20].

Among the nature-inspired swarm optimization algorithms, a noteworthy algorithm presenting novel ideas and a powerful optimization scheme is the Cheetah Optimizer (CO) proposed by Akbari et al. [21]. This swarm algorithm is inspired by the movements and behaviors adopted by a group of cheetahs during the process of pursuit and attack of a prey, simulating these behaviors in a novel competitive algorithm with multiple collective skills capable of locating the global optimal solution requiring a minor number of operations and evaluations, making this algorithm a powerful and efficient alternative applied to diverse engineering applications, such as the regulation and control of parameters for photovoltaic cells [22, 23] or the management and planning of trajectories for Unmanned Aerial Vehicles (UAVs) [23].

Nature-inspired metaheuristic algorithms represent an easy and competitive alternative for solving complex optimization problems thanks to their simplicity in their operations and modeling of complex phenomena with easily interpretable mathematical operators. Nevertheless, several of these algorithms tend to present unbalanced exploration and exploitation stages, considerably reducing their performance. Metaheuristic algorithms lack a proper modulation of the diversity of the population, which severely affects generating clusters of individuals in determined regions and revisiting zones with poor quality. The imbalance in exploration and exploitation has brought the hybridization of these algorithms [24], aiming to overcome these difficulties and gradually improve the capabilities of the original optimization schemes [25].

For all metaheuristic algorithms, explorative and refinement properties play a fundamental role in the finding of high.-quality solutions in a highly efficient way [26]. A commonly used technique for considering more zones of the search space is Opposition-Based Learning (OBL), which has evidenced interesting advantages for the exploration by considering the concept of the opposition by exploring regions of the space where no individual has explored yet, discovering new potential solutions. On the other hand, diversity control also considerably influences how an optimization algorithm can regulate its exploration and exploitation of solutions [27]. An interesting method for estimating the existing diversity in a population is the one proposed by Morales-Castañeda et al. [28], which also is employed as a moderator based on the diversity at certain stages of the algorithm, generating new movements on the individuals or generating new distributions on the positions of the candidate solutions to keep a more controlled balance on the exploration and the exploitation.

This chapter presents a novel hybridization of the Cheetah Optimizer employing the Opposition-Based Learning methodology and a technique to control the diversity of the population. This new hybrid approach considers two different positions with independent movements to keep a better knowledge of the searching space. At the same time, diversity control modulates the exploration–exploitation balance depending on evolutionary progress, adjusting different strategies to achieve better dispersion or concentration of the candidate solutions. For a correct evaluation of this approach, exhaustive experimentation has been performed against the performance of the original Cheetah Optimizer and other metaheuristic algorithms from the literature employing a set of 30 traditional benchmarking test functions, including multimodal, unimodal, hybrid and shifted functions to evaluate the capabilities to find the global solutions even in displaced spaces. Experimental results demonstrate enhanced explorative and refinement capabilities compared to the original algorithm.

This chapter is organized as follows: Sect. 6.2 analyzes and explains the operations of the original Cheetah Optimizer. Section 6.3 discusses the methodologies employed in hybridization, which are the OBL and diversity measure control. Section 6.4 discusses the implementation of the methodologies in the hybrid approach. Section 6.5 discusses the experimental results. Finally, Sect. 6.6 discusses the conclusions.

6.2 Original Cheetah Optimizer

The Cheetah Optimizer (CO) [21], originally proposed by Akbari et al., is presented as a novel swarm metaheuristic algorithm inspired by the diverse behaviors and movements adopted in a group of wild cheetahs during a hunting process of prey in their natural habitat. The algorithm combines three different strategies based on three different movements adopted during the hunt: the cheetahs move freely along the space searching for potential prey, vigilate the movements of a detected prey expecting a more vulnerable position of the prey and attack the prey directly at maximum speed. In addition to the mechanisms adopted, the algorithm also incorporates a "leave prey and go back home" mechanism as an auxiliary mechanism to improve the diversity in the individuals, the convergence rate of the solutions and the robustness of the individuals, dynamically avoiding premature convergences.

6.2.1 Definition of the Operators

In the Cheetah Optimizer (CO), the population $\mathbf{X}\{\mathbf{x}_1, \mathbf{x}_2, \ldots, \mathbf{x}_n\}$ of n cheetahs are constantly updated from the position given in the initialization until the stop criterion is reached, which may be a determined number of iterations or evaluations in the cost function. Each cheetah $\mathbf{x}_{i \in n}$ corresponds to an d-dimensional array $\{x_{i,1}, x_{i,2}, \ldots, x_{i,d}\}$. To determine the quality of the cheetahs, the cost function $J(x)$ is used to determine the quality of the solution for each $\mathbf{x}_i^k$ Since each cheetah is evaluated constantly, the position of the prey is constantly updated as the global solution is updated, given that more promising solutions are detected. The update of each cheetah may follow three different movements according to the time elapsed of hunting and their positions with respect to the prey (the global solution so far). The following rules dictate the movements a cheetah can adopt:

- *Searching:* the cheetahs are exploring the search space, evaluating their positions and looking for a candidate prey (promising solutions).
- *Sitting-and.waiting:* When a prey is detected in space, cheetahs adopt a passive stance, allowing the prey to move freely to a more feasible position, leaving it more vulnerable by being closer to the position of the cheetah.
- *Attacking:* In this mechanism, the cheetahs move directly to the prey at high speed, allowing flexibility in the movements of the cheetahs to prevent a possible runaway from the prey.

Complementary to the movement mechanisms, when the cheetahs do not reach the prey properly or the hunting time has been exceeded, the cheetahs may adopt a *leave the prey and go back home* when the cheetahs may return to their initial positions and restart their hunting.

6.2.2 Population Initialization

The initial step for all metaheuristic algorithms is to define the initial position of the candidate solutions across the search space defined for the optimization problems. This also helps the computational algorithms to trace an efficient strategy based on the positions and the initial quality of the solutions, getting a certain knowledge of the problem and deciding which direction to follow. The original Cheetah Optimizer (CO) follows a random initialization, where the j decision variables for all the i members of the population are defined by following a pseudo-random uniform distribution along the lower $\boldsymbol{P}_{lb,j}$ and upper $\boldsymbol{P}_{ub,j}$ boundaries of the search space. The initialization methodology employed by the CO algorithm follows the Equation:

$$\mathbf{x}_{i(1,\ldots,n),j(1,\ldots,d)} = \boldsymbol{P}_{lb,j} + U(0,1) \cdot \left(\boldsymbol{P}_{ub,j} - \boldsymbol{P}_{lb,j}\right) \tag{6.1}$$

6.2.3 Selection of Active Members

The first step in the CO is to select the individuals in the population who will actively participate in the hunting, updating their positions by the selection of strategies. In this selection, a random subset of individuals is selected, limited by a variable m, which determines the number of individuals of the population selected randomly. The number of individuals is selected by the constraint $2 \leq m \leq n.$. Selecting a portion of the population enhances the dynamism in the movements for selecting new positions and obtaining diversity in the energy levels of the individuals (cheetahs).

6.2.4 Selection of Strategies

Once all the active members for the update cycle are selected, the next step is to determine the variables that establish the movement adopted for each cheetah. All the individuals may adopt different strategies independently; this means that even if a cheetah adopts a searching mechanism, others may take an attacking strategy or the sitting-and-waiting mechanism. However, the selection of strategies may be influenced by their position regarding prey or elapsed hunting time. To select a proper strategy for each individual, two random variables r_2 and r_3 intervene in the selection of a strategy; if $r_2 > r_3$, the cheetahs adopt a sit-and-wait mechanism; otherwise, the cheetahs may adopt an active movement; however, the algorithm tends to consider the selection of the searching strategy with greater weight. For selecting a different movement, the selection criterion is determined following Eq. (6.2):

$$H = e^{2\left(1-\frac{t}{T}\right)}(2r_1 - 1) \tag{6.2}$$

In Eq. (6.1), H corresponds to a factor for selecting the strategy, t corresponds to the current hunting time so far, T corresponds to the maximum length of the hunting process, which is determined as $T = 60 \times \frac{d}{10}$, and r_1 represents a random, uniformly distributed number from 0 to 1. Based on the hunting time, the algorithm gives more weight to determined mechanisms as the refinement of feasible solutions or exploration of the search space. This H factor plays a crucial role in selecting which movement an individual may adopt at a certain time of the evolutionary process. The H also plays an important role in the exploration–exploitation balance, deciding whether an explorative (searching across the searching space for promising regions) and refinement (attacking the best-detected solutions) may intervene at a determined time. The adjustment of these parameters aids in preventing local optimal solutions and improves the ability to locate the global optimal solution easily.

6.2.5 Searching Mechanism

The searching mechanism focuses on an active movement of the individuals, exploring the environment and looking for a promising prey in the space. If the prey stays in movement, the finding of feasible solutions becomes easier. The prey is denoted as the global best-so-far of the population, labeled as $\mathbf{x}_{Best}$, meanwhile the rest of the population $\mathbf{x}_i\left\{x_{i,1}, x_{i,2}, \ldots, x_{i,d}\right\}$ are the active cheetahs of the population. In this searching movement, the cheetahs adopt a random movement determined by Eq. (6.3):

$$\mathbf{x}_{i,j}^{t+1} = \mathbf{x}_{i,j}^{t} + \hat{r}_{i,j}^{-1} \cdot \alpha_{i,j}^{t} \tag{6.3}$$

where $\mathbf{x}_{i,j}^{t+1}$ corresponds to the new solution and $\mathbf{x}_{i,j}^{t}$ corresponds to the current selected candidate solution, $\hat{r}_{i,j}^{-1}$ corresponds to a randomization factor represented

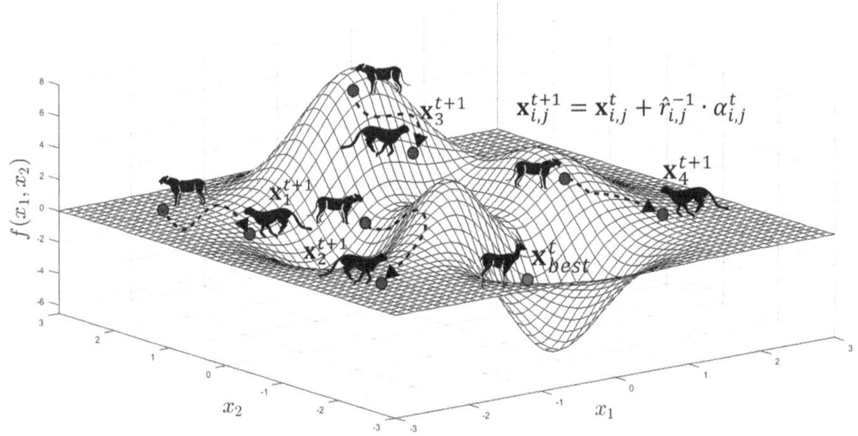

Fig. 6.1 Graphical representation of the searching mechanism of the CO

as a random number from a normalized distribution, and $\alpha_{i,j}^t$ corresponds to the step length for the moving cheetah. This movement may differ for a regular hunting cheetah and for the leader, who is an individual closer to the prey. The $\alpha_{i,j}^t$ is different for each individual and is calculated following Eq. (6.4).

$$\alpha_{i,j}^t = \begin{cases} 0.001 \cdot \frac{t}{T} \cdot \left(P_{ub,j} - P_{lb,j}\right), \textit{leader} \\ 0.001 \cdot \frac{t}{T} \cdot \left(\mathbf{x}_{r,j}^t - \mathbf{x}_{i,j}^t\right), \textit{other member} \end{cases} \tag{6.4}$$

In this criterion, $\mathbf{x}_{r,j}^t$ is another individual of the population selected randomly, and $P_{ub,j}$ and $P_{lb,j}$ correspond to the upper and lower boundaries of the search space, respectively. For a better understanding of this method, Fig. 6.1 presents a graphical representation of the movements of a group of cheetahs in a 2-D search space delimitated by the boundaries $-3.2 \leq x \leq 3.2$ for a objective function defined in Eq. (6.5).

$$f(x, y) = 3(1-x)^2 + 10\left(\frac{x}{5} - x^3 - y^5\right)e^{\left(-x^2-y^2\right)} - \frac{1}{3}e^{\left(-(x+1)^2-y^2\right)} \tag{6.5}$$

6.2.6 Sitting-and-Waiting Mechanism

This is another of the mechanisms the algorithm can adopt for the update of a determined cheetah. In this mechanism, the cheetahs adopt a passive behavior, staying in their current positions but inspecting the movements of the prey, expecting it to approach the hunter, placing itself in a vulnerable position. In this case, as the cheetah remains in its current position, the new position is determined in Eq. (6.6).

$$\mathbf{x}_{i,j}^{t+1} = \mathbf{x}_{i,j}^{t} \tag{6.6}$$

This mechanism allows the prey (the global best solution) to move freely. Still, the hunters keep a trace of the movements and the positions, aiming to reach this prey in a more compromising position (the quality of this solution is improved, or the position of the current solution finds a more promising quality).

6.2.7 Attacking Mechanism

This mechanism considers two different aspects in the movement of the individuals to get a higher success ratio in a direct attack: the movement speed of a hunter to attack a prey and the agility to switch its movement direction to avoid the prey running away and escaping from its predators. In real groups of cheetahs, when a prey detects the danger, it tries to make evasive movements to escape and move to a safe position. The Cheetahs use fast and flexible movements to efficiently approach the prey, based on the position and the distance existing between a hunter and the prey. The Cheetah Optimizer (CO) algorithm models the movement of the i cheetahs for the j array of decision variables defined in Eq. (6.7):

$$\mathbf{x}_{i,j}^{t+1} = \mathbf{x}_{best,j}^{t} + \check{r}_{i,j} + \beta_{i,j}^{t} \tag{6.7}$$

In the movement equation described in Eq. (6.5), $\mathbf{x}_{best,j}^{t}$ corresponds to the prey (the global best-so-far individual in the population or the individual with the best fitness in the population), $\beta_{i,j}^{t}$ does the difference between the current define an interaction factor $\mathbf{x}_{i,j}^{t}$ cheetah and another member of the population or even the leader, defined by the Eq. (6.8):

$$\beta_{i,j}^{t} = \mathbf{x}_{r,j} - \mathbf{x}_{i,j} \tag{6.8}$$

And $\check{r}_{i,j}$ is a movement factor that can define the movement direction of the cheetah, which is defined by:

$$\check{r}_{i,j} = \left|r_{i,j}\right|^{\exp\left(\frac{r_{i,j}}{2}\right)} \sin(2\pi r_{i,j}) \tag{6.9}$$

In Eq. (6.9), $r_{i,j}$ is a random, normally distributed number. This equation models an attacking model for the cheetahs (candidate solutions) where speed plays a crucial role in the movement. Still, the cheetahs may also adopt different directions to surround the prey and get a better success rate in the hunting process. Following the movement described in Eq. (6.7), Fig. 6.2 exposes a graphical representation of the attacking movement adopted by the Cheetahs for a searching space defined in Eq. (6.5) constrained by $-3.2 \leq x \leq 3.2$.

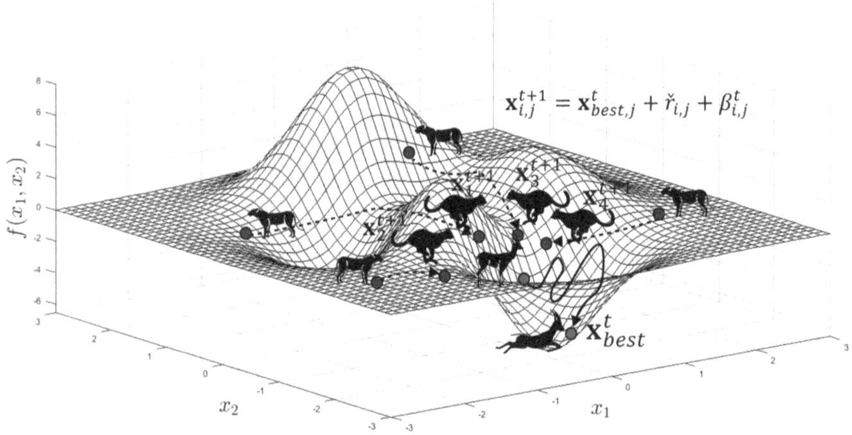

Fig. 6.2 Graphical representation of the attacking mechanism of the CO

6.2.8 Leave the Prey and Go Back Home

All the cheetahs have different energy levels for each different movements and to get a proper hunting mechanism. Such levels are modeled by a hunting timer that constantly increases as the evolutionary, iterative process of the algorithm passes by. When the hunting timer has reached the limit, and the hunters are not capable of reaching the prey, a group of cheetahs leaves their current positions and return to home (the original authors define the home as the initial positions), aiming to avoid the individual get stuck in suboptimal regions and enhance the diversity of the algorithm.

6.2.9 General Procedure

For a better and more detailed understanding of the general procedure of the Cheetah Optimizer, this section presents a pseudocode where the processes are explained in sequential order in Algorithm 1. First, in Line 1, the initial parameters are defined for the algorithm, establishing the population size (n), the dimensionality (d), the boundaries for the optimization problem (P_{ub}, P_{lb}), the maximum number of iterations for the algorithm ($MaxIt$), and also initialize the hunting timer (t) and the maximum hinting time (T). Lines 2–4 establish the initialization process defined in the Eq. (6.1), set the home (initial positions), and also find the positions of the leader and the prey who will play a vital role in the movements of the cheetahs. From Line 5, the iterative process, first by selecting the m active members for that iteration and selecting the neighboring $\mathbf{x}_r$ individual. Line 9 defines the update process for the $j\{1, \ldots, d\}$ decision variables, calculating the $\check{r}$, $\hat{r}$, α (Eq. (6.4)), β (Eq. (6.8)) and

H (Eq. (6.2)) parameters. Lines 11–13 establish the random factors r_2, r_3 and r_4 employed for the selection of mechanisms between the searching (Eq. (6.3)), sitting and waiting (Eq. (6.6)) or attacking (6.7). Once all the dimensions have been updated following different strategies, the boundaries are validated, avoiding the new solutions that are not outside of the limits, and in Line 25, the new prevailing solution is updated with the fittest individual, increasing by one the hunting timer t. Finally, in Line 26, if a certain time in the hunting process has passed and the position of the leader has not presented any variation in a determined time, the hunting is considered unsuccessful, and the cheetahs return to their initial positions.

Algorithm 1. General procedure of the Cheetah Optimizer (CO)

1. Define n, d, P_{ub}, P_{lb} $MaxIt$, $t \leftarrow 0, T \leftarrow 60 \times \frac{d}{10}$
2. $\mathbf{X}(\mathbf{x}_1, \ldots, \mathbf{x}_n) \leftarrow RandomInitialization(P_{ub}, P_{lb})$
3. Save initial positions for $\mathbf{X}$
4. Set the prey $\mathbf{x}_{Best}$ and the leader positions
5. for $it \in MaxIt$
6.   Select $m(2 \leq m \leq n)$ individuals randomly
7.   for $i \in m$
8.   Select the $\mathbf{x}_r$ neighbor for the $\mathbf{x}_i^t$
9.   for $j \in \{1, \ldots, d\}$
10.   Calculate $\check{r}$, $\hat{r}$, α, β and H
11. $r_2, r_3 \leftarrow U(0,1)$
12.   if $r_2 \leq r_3$
13. $r_4 \leftarrow U(0,3)$
14.   if $H \geq r_4$
15. $\mathbf{x}_{new}^t \leftarrow AttackingStrategy$
16.   else
17. $\mathbf{x}_{new}^t \leftarrow SearchingStrategy$
18.   end if
19.   else
20. $\mathbf{x}_{new}^t \leftarrow SittingAndWaitingStrategy$
21.   end if
22.   end for
23. $t \leftarrow t + 1$
24.   Verify boundaries for $\mathbf{x}_{new}^t$
25.   Update the new $\mathbf{x}_i^{t+1}$ solution with $\mathbf{x}_{new}^t$ if $f(\mathbf{x}_{new}^t) \leq f(\mathbf{x}_i^t)$
26.   if $t > rand \times T$ and the leader has not moved in a time
27.   Apply $LeavePreyAndGoBackHome$ and change the leader
28.   Replace the prey position by $\mathbf{x}_i$
29.   end if
30.   Update the prey ($\mathbf{x}_{Best}$)
31. end for

6.3 Employed Methodologies

In this section, two methodologies employed in the proposed hybrid approach are explained and analyzed: The Opposition-Based Learning (OBL) [29], which is a commonly used technique in metaheuristic optimization to enhance the exploration of high-scale exploration spaces by considering diverse positions simultaneously and increasing the probability to find promising regions by taking into consideration unexplored areas, and the importance of the diversity quantitative analysis and control for the adaptation of new strategies to separate the individuals or to focus in refining a different area [28]. Keeping a balance between the exploration and exploitation stages is fundamental to getting an efficient search strategy, avoiding suboptimal regions, and finding the global optimal solution in a minor time and with a minor number of computational resources easily.

6.3.1 Opposition-Based Learning (OBL)

The main concept of Opposition-Based Learning [29] was originally analyzed by Tizhoosh in 2005 [30] as a powerful paradigm in the machine learning methodologies to increase the knowledge of a determined problem, strongly used in different machine learning schemes, such as the reinforcement learning, neural networks and optimization algorithms. The learning process for the computational intelligence systems is commonly given by random points following diverse steps. Not all intelligence algorithms follow a linear step to find a solution. Still, they apply a learning process starting from an arbitrary point and constantly learning about its environment, for example, the weights for reasoning from a neural network, the population positions in an optimization algorithm or the parameters and policies for a reinforcement learning algorithm. Computational artificial intelligence mechanisms do not know the optimal solutions from the beginning; they must explore and discover from a wide spectrum of solutions until the best-quality objective is found. However, the process of finding promising solutions by random steps may take a considerable time, even when it is assumed that the current position is very far from the global best solution. Oppositon-Based Learning makes the process of finding promising solutions easier by proposing the consideration of multiple directions in the space; in this case, when a position is considered following a random guess, another position is taken into account by the opposite part in the space.

For multidimensional spaces, Tizhoosh defined the OBL process as: "*For a n-dimensional array of values $P(x_1, x_2, \ldots x_n)$ where all the points $x_1, \ldots, x_n \in \Re$ subject to the boundaries $x_i \epsilon [a_i, b_i]$. The opposite number $\tilde{x}$ is defined as:*"

$$\widetilde{x_i} = a_i + b_i - x_i, i = 1, \ldots, n \tag{6.10}$$

Once the new opposite guess is defined for a determined value in the problem, both new candidates are modified following a determined update rule dictated in the algorithm. This gives two results $f(x)$ and $f(\tilde{x})$. Finally, it is determined which guess presents a better quality to the optimization problem. In other words, if $f(\tilde{x}) \leq f(x)$, the new solution for the problem is replaced by $\tilde{x}$, otherwise, the process continues with the x.

6.3.2 Diversity Control

A key factor for proper performance on a metaheuristic algorithm is the correct balance between intensification and diversification [31]. Diversification refers to the generation of new solutions along the search space, continuously exploring and obtaining new information about the problem to decide whether a region is worthy or not to focus efforts on refining. On the other hand, intensification is the focus of the efforts to exploit and refine the obtained in a determined region where high-quañlity solutions have been reported.

Despite not all metaheuristic algorithms being able to find the best global solution for all the optimization problems, the main target is to get an algorithm that presents high-quality solutions in a minor time [32]. Since all the optimization problems applied in real life present several constraints and a high complexity, a correct balance between the explorative and refinement capabilities is crucial in the performance of an algorithm to learn from the problem efficiently, avoid a premature convergence and prevent the stagnation in zones with a poor quality ignoring the finding of the global best solution.

Morales Castañeda et al. proposed a quantitative method to analyze the behaviors detected along the process of a metaheuristic algorithm and analyze the exploration and exploitation patterns and the transition in these steps [28]. For reaching a proper balance on the exploration and exploitation percentage presented in different stages of the evolutionary process, this analysis utilizes the existing distances between all the members of the population to determine how the individuals explore the search space or exploit promising solutions; when the metaheuristic algorithm adopts an exploitation mechanism, the individuals tend to group in a determined point decreasing the distance among the individuals and increasing the exploitation efforts. Conversely, an increase in the distance indicates that the individuals are separating from themselves, increasing the exploration efforts.

The calculation of the distances existing between all the individuals of the population is determined by a dimension-wise diversity measurement, where the diversity is defined as:

$$Div_j = \frac{1}{n}\sum_{i=1}^{n}\left|median\left(x^j\right) - x_i^j\right| \tag{6.11}$$

$$Div = \frac{1}{m}\sum_{j=1}^{m} Div_j \tag{6.12}$$

In Eq. (6.11), $median\left(x^j\right)$ represents the median of the j-th decision variable for all the elements of the population and x_i^j corresponds to the j-th decision of the current

x_i search agent. The calculation of diversity is first measured for each independent decision variable of all the search agents. Finally, the total diversity of the individuals is summarized and calculated by Eq. (6.12).

To get a proper balance between the exploration and exploitation stages, this methodology calculated the exploration and exploitation percentages at different times of the algorithm. As the evolutionary process progresses along the entire algorithm process, it is expected that the exploration will gradually decrease and the exploitation will constantly increase, making a proper transition between the explorative and refining patterns. The measurement of the exploration *XPL* and the exploitation *XPT* percentages is defined by:

$$XPL = \left(\frac{Div}{Div_{Max}}\right) \times 100 \tag{6.13}$$

$$XPT = \left(\frac{|Div - Div_{Max}|}{Div_{Max}}\right) \times 100 \tag{6.14}$$

In Eqs. (6.13) and (6.14), Div_{Max} is the maximum diversity observed along all the evolutionary process. The *XPT* percentage is a complementary calculation for the *XPL* percentage, given the difference between the maximum diversity and the current diversity at the current iteration.

6.4 The Proposed Hybrid Methodology

The CO algorithm is an efficient and powerful metaheuristic methodology for solving high-complexity optimization problems thanks to the combination of different strategies that prioritize both exploration and exploitation based on the progress of the optimization process and the finding of promising solutions. This chapter is intended to expose a novel hybrid approach of the original CO integrating into the optimization process the OBL and the incorporation of a more controlled diversity control based on the progress of the algorithm. The integration of these methods overcomes possible weaknesses of the algorithm and gradually potentializes the strong points of this algorithm.

6.4.1 Update Behaviors Applying the OBL

In this hybrid approach, the update rules (or hunting mechanisms) employ Opposition-Based Learning (OBL) with the objective of acquiring new important information about the search space to enhance the algorithm's decision-making process. The CO algorithm selects a number of $m \in n$ search agents to participate in the hunting process for an iteration actively, alternating between the searching, sitting-and-waiting and attacking mechanisms for all the j dimensions of each individual.

For all the members who participate in the hunting process actively, the OBL approach generates a new candidate solution directly in the opposite position of the current position for all the j dimensions. This new OBL candidate solution is determined by taking into account the lower $P_{lb,j}$ and upper $P_{ub,j}$ boundaries of the search space following the Eq. (6.15):

$$\mathbf{x}^{t}_{OBL,j} = P_{lb,j} + P_{ub,j} - \mathbf{x}^{t}_{i,j} \tag{6.15}$$

This new individual generated by the OBL follows the same movement mechanism adopted by the current $\mathbf{x}_i$ individual and generates a new solution; however, since this individual presents a different distance to the leader of the prey, this new OBL individual considers different parameters calculated independently according to the selected movement.

6.4.2 Searching Strategy on the OBL Individual

In the search movement determined by the Eq. (6.2) defines the movement by the $\hat{r}$ random factor and the α step length parameter, defined differently for the leader and the other cheetahs of the population. When the searching strategy is selected for the current $\mathbf{x}_i$ individual, the $\mathbf{x}_{OBL}$ will follow the same movement mechanisms, following a different pattern by the position of this individual, and considering different randomness factors that influence its movement, direction and magnitude. For the $\mathbf{x}_{OBL}$ candidate individual, the movement is defined as:

$$\mathbf{x}^{t}_{new_{OBL},j} = \mathbf{x}^{t}_{OBL,j} + \hat{r}^{-1}_{OBL,j} \cdot \alpha^{t}_{OBL,j} \tag{6.16}$$

In Eq. (6.16), $\hat{r}^{-1}_{OBL,j}$ is a different random number from a uniform distribution from 0 to 1, with a different value from $\hat{r}^{-1}_{i,j}$, and the $\alpha^{t}_{OBL,j}$ is the step length for the OBL individual considering its new distance to the neighbor individual $\mathbf{x}_{r,j}$ selected for the $\mathbf{x}_i$ individual. For this new individual, this parameter is calculated following Eq. (6.17).

$$\alpha^{t}_{OBL,j} = 0.001 \cdot \frac{t}{T} \cdot \left(\mathbf{x}^{t}_{r,j} - \mathbf{x}^{t}_{OBL,j}\right) \tag{6.17}$$

6.4.3 Attacking Strategy for the OBL Individual and Selection

When the hunting mechanism is selected for the current update rule for the $\mathbf{x}_i$ individual, the $\mathbf{x}_{OBL}$ will follow simultaneously the same mechanism. In the attacking mechanism defined in Eq. (6.7), the $\beta^{t}_{i,j}$ factor is defined by the difference existing for all the j decision variables between the current individual and another individual randomly selected from the population and $\check{r}_{i,j}$ defined by Eq. (6.9) determines the direction of the movement, allowing fast and flexible movements by the hunter in the attacking mode. In this hybrid approach, the $\check{r}_{i,j}$ keeps the same value for both $\mathbf{x}_{i,j}$ and $\mathbf{x}_{OBL,j}$ individuals, consdidering a different movement parameter calculated separately for the current and the OBL individuals, obtaining $\beta_{i,j}^{t}$ and $\beta_{OBL,j}^{t}$, considering now the existing difference for all the j dimensions between the current $\mathbf{x}_i$ and $\mathbf{x}_{OBL}$ regarding the prey $\mathbf{x}_{Best}$ following the Eqs. (6.18) and (6.19):

$$\beta_{i,j}^{t} = \mathbf{x}_{best,j} - \mathbf{x}_{i,j} \tag{6.18}$$

$$\beta_{OBL,j}^{t} = \mathbf{x}_{best,j} - \mathbf{x}_{OBL,j} \tag{6.19}$$

Once the current and the OBL cheetahs have been updated following the selected movement mechanisms, only the individual with the best fitness will prevail as the new cheetah for the next generation. In this case, given the fitness of the updated individual $\mathbf{x}_{\boldsymbol{new}}^{t}$ and $\mathbf{x}_{new_{OBL}}^{t}$, both are evaluated in the cost function, remaining only with the cheetah with the fittest solution:

$$\mathbf{x}_{i}^{t+1} = \begin{cases} \mathbf{x}_{new_{OBL},j}^{t}, \; iff\left(\mathbf{x}_{new_{OBL}}^{t}\right) \leq f\left(\mathbf{x}_{\boldsymbol{new}}^{t}\right) \\ \mathbf{x}_{new}^{t}, \; otherwise \end{cases} \tag{6.20}$$

6.4.4 Diversity Control Mechanism

The proposed scheme considers the diversity of the population as a key factor in the dynamic movements and is auxiliary to prevent getting trapped in determined locations. In this hybrid approach, better diversity control management is obtained by applying a modification of the original "leave the prey and go back home" mechanism of the original CO, applying the diversity management quantitative analysis [28]. The exploration (*XPL*) and exploitation (*XPT*) percentages previously defined in Eqs. (6.13) and (6.14), respectively, indicate the progress of the evolution of the candidate solutions, evaluating the expected capabilities of the algorithm to explore the search space at early stages of the optimization process, and the capability of the solutions to converge at the optimal zone at later stages of the algorithm. Preventing premature convergence is crucial in the exploration of the algorithm by increasing the likelihood of finding higher-quality regions and locating the global optimum easily.

In the proposed hybrid approach, the diversity control considers an exploration tolerance of approximately 15% along the early steps of the algorithm, expecting behaviors more focused on exploring the search space, looking for different solutions and identifying promising solutions in the space. Suppose the *XPL* percentage presents a value lower than 15% in the first half of the algorithm. In that case, the algorithm employs a modified "leave prey and go back home" mechanism with OBL, selecting a number of individuals of the population of the population and locating

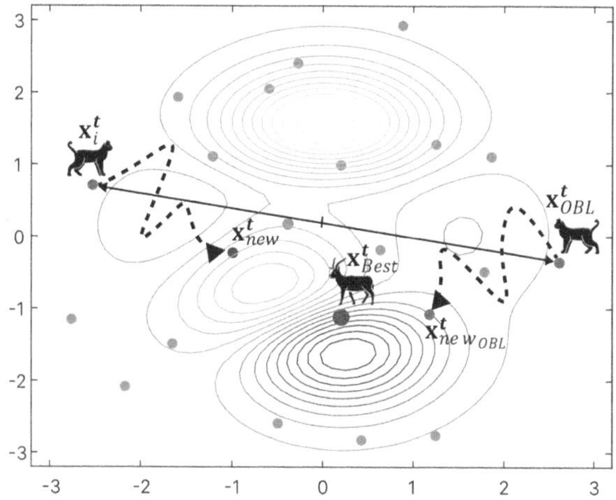

Fig. 6.3 Graphical representation of the OBL movement determined by the hybrid cheetah optimizer approach

them in the opposite part of their original positions. The application of the OBL on this relocation of individuals brings not only more controlled diversity and prevention of a possible premature convergence but also new information about unexplored areas and better diversity in the search mechanisms.

In Fig. 6.3, a graphical representation of the movement adopted by a current $\mathbf{x}_i$ and an $\mathbf{x}_{OBL}$ individual is represented. Tough both individuals follow the same mechanisms, considering their different positions by the opposite individual and considering the distance between the global best-so-far and the individuals, both candidate solutions get different movement patterns in different positions. Once both individuals adopt their new positions, only the one with the fittest solution will proliferate for the next generation.

6.4.5 General Procedure of Hybrid Approach

For a better understanding of the general procedure of the hybrid Cheetah Optimizer with Diversity and Opposition Learning, named CO-DO, in Algorithm 2, a summary of the algorithm with the integration of the OBL and the diversity measurement is exposed. The hybrid approach keeps several features and strengths of the original CO. In Line 9, once the current individual has been chosen and the neighboring individual is also selected, the $\mathbf{x}_{OBL}$ individual is generated by applying the OBL scheme from Eq. (6.15). Additionally, the parameters for the movement behaviors applied for the OBL individual are calculated simultaneously as the α_{OBL} by following Eqs. (6.17), the new parameter β_i by following the Eq. (6.18) and β_{OBL} following Eq. (6.19). The movement behaviors of Lines 16, 19 and 23 also update the OBL individual simultaneously. In Line 28, the validation of the values inside of the boundaries is also performed for both new and OBL individuals obtained by the movement rules. In Line 29, before the update of the $\mathbf{x}_i^{t+1}$, the $\mathbf{x}_{new}^t$ new individual is updated according to the fitness between the $\mathbf{x}_{new}^{t}$ and the $\mathbf{x}_{new_{OBL}}^{t}$, where if if $f\left(\mathbf{x}_{new_{OBL}}^{t}\right) \leq f\left(\mathbf{x}_{new}^{t}\right)$, then the solution obtained by the OBL process is selected as the new candidate solution for the next generation. Otherwise, the solution obtained by the update of the current $\mathbf{x}_i$. In Line 32 of the Algorithm 2, employing the diversity calculations explained in the Subsection 6.3.2 and following the Eqs. (6.11) and (6.12), the exploration and exploitation percentages XPL, XPT are calculated the Eqs. (6.13) and (6.14). Once the exploration and exploitation percentages are calculated, a comparison is performed to assess the exploration percentage at the first half of the process of the algorithm. If a low exploration percentage is detected during the first step of the process, the modified "leave prey and go back home" mechanism is adopted. In this strategy, first, the population is sorted in terms of their fitness values from the best to the worst. From this sorted population, a number p indexes of the worst element on this sorted population is selected, which in this proposal is proposed as 30% of the total population size. This selected array of p indexes is the individual that will return to the opposite part of their original place, considering new information of the space and getting more diverse environments.

Algorithm 2. General procedure of the Cheetah Optimizer with Diversity and Opposition Learning (CO-DO)

1. Define n, d, P_{ub}, P_{lb} $MaxIt, t \leftarrow 0, T \leftarrow 60 \times \frac{d}{10}$
2. $\mathbf{X}(\mathbf{x}_1, \ldots, \mathbf{x}_n) \leftarrow RandomInitialization(P_{ub}, P_{lb})$
3. Save initial positions for $\mathbf{X}$
4. Set the prey $\mathbf{x}_{Best}$ and the leader positions
5. for $it \in MaxIt$
6.   Select $m (2 \leq m \leq n)$ individuals randomly
7.   for $i \in m$
8.   Select the $\mathbf{x}_r^t$ neighbor for the $\mathbf{x}_i$
9.   Obtain the OBL individual $\mathbf{x}_{OBL}$ (Equation (15))
10.   for $j \in \{1, \ldots, d\}$
11.   Calculate $\check{r}_i, \hat{r}_i, \alpha_i, \beta_i, \check{r}_{OBL}, \hat{r}_{OBL}, \alpha_{OBL}, \beta_{OBL}$ and H
12. $r_2, r_3 \leftarrow U(0,1)$
13.   if $r_2 \leq r_3$
14. $r_4 \leftarrow U(0,3)$
15.   if $H \geq r_4$
16. $\mathbf{x}_{new}^{t} \leftarrow AttackingStrategy(\check{r}_i, \beta_i)$
17. $\mathbf{x}_{new_{OBL}}^{t} \leftarrow AttackingStrategy(\check{r}_{OBL}, \beta_{OBL})$
18.   else
19. $\mathbf{x}_{new}^{t} \leftarrow SearchingStrategy(\hat{r}_i, \alpha_i)$
20. $\mathbf{x}_{new_{OBL}}^{t} \leftarrow SearchingStrategy(\hat{r}_{OBL}, \alpha_{OBL})$
21.   end if
22.   else
23. $\mathbf{x}_{new}^{t} \leftarrow SittingAndWaitingStrategy$

```
            x^t_newOBL ← SittingAndWaitingStrategy
        end if
    end for
    t ← t + 1
    Verify boundaries for x^t_new and x^t_newOBL
    Update the x^t_new with x^t_newOBL if f(x^t_newOBL) ≤ f(x^t_new)
    Update the new x_i^{t+1} solution with x^t_new if f(x^t_new) ≤ f(x_i^t)
    Calculate diversity Div by Equation (12)
    Calculate XPL (Equation (13)) and XPT (Equation (14))
    if XPL < 15 and it ≤ MaxIt × 0.5
        hierarchy ← Sort(X)
        p ← hierarchy((n × 0.7):n)
        for k = p
            x_k ← GoBackHomeWithOBL
        end for
    end if
end for
```

6.5 Experimental Results

In this section, the results obtained for the experimental basis of the proposed approach are analyzed and discussed. The proposed Cheetah Optimizer with Diversity and Opposition-Learning (CO-DO) against the performance of the original Cheetah Optimizer (CO) and seven well-established metaheuristic approaches from the literature selected by their notorious features for both exploration and exploitation and applicability for real-life complex optimization problems. The selected algorithms for the evaluation include the Artificial Bee Colony (ABC) [12], the Cuckoo Search [14], the Harmony Search [18], the Bat Algorithm (BA) [13], the States of Matter Search (SMS) [19], the Particle Swarm Optimization (PSO) [9] and the Differential Evolution (DE) [6]. An assessment of these approaches has been conducted through a set of 28 diverse functions from the traditional benchmarking test functions set. [33], including 14 multimodal, five unimodal, three hybrid and six shifted functions. The selection of these functions for the Comparison is due to the combination of different levels of complexity, including the combination of multiple local optimal solutions and the combination of functions with displaced global optimal solutions, with the main objective of testing the capabilities of the algorithms to search thoroughly along the space and avoid the attraction to certain positions on the algorithms, for example, attraction to the origin point. Table 6.9 from the Appendix summarizes all the employed test functions and includes a mathematical representation. All of the experimentation was performed in a computer equipped with a 64-bit Windows

Table 6.1 Employed parameters for all the algorithms

Algorithm	References	Set parameters for the algorithm
ABC	[12]	Limit $L = 50$
CS	[14]	Finding alien eggs $p_a = 0.25$
HS	[18]	$HMS = 50$, $HMCR = 0.7$, $PARate = 0.3$
BA	[13]	$A = 0.9$, $r = 0.2$, $Q_{min} = 0$, $Q_{max=1}$
SMS	[19]	$\alpha = [0.8, 0.4, 0.1]$, $\beta = [1.0, 0.7, 0.5]$, $p = [0.9, 0.2, 0.0]$, $\rho = [0.8, 1.0, 0.3, 0.6, 0.0, 0.1]$
PSO	[9]	$c_1 = 2$, $c_2 = 2$, $w = 0.9$, $w_{max} = 0.9$, $w_{min} = 0.2$
DE	[6]	$pCr = 0.9$, $F = 0.2$
CO	[21]	Active members $m = 17$

10 Por operating system, Intel® Core™ i7-4700 processor with 16 GB of RAM and 1 TB of solid statute disk (SSD) performed with MATLAB R2022b software.

For all the algorithms employed for the Comparison, a set of 30 independent executions for each test function generates a total of 30 different outputs per algorithm, considering a population size n of 50 search individuals. The parameters employed for the optimal results of the algorithms are explained in Table 3.1. It is worth noting that the CO employs a m number of active members of 17. For a fair comparison between the original methodology, the CO-DO considers the same number of m active members, corresponding to a third part of the total population (Table 6.1).

The analysis of the performance considers three different factors for the set of the 30 independent executions of the algorithms, each of them representing different aspects of the performance of an algorithm: The Average Best (AB), the Median (MD) and the Standard Deviation (SD). Since the experimentation gives a total of 30 independent results per algorithm and test function, these results are averaged to obtain the AB. Meanwhile, the MD is the median of these 30 results, and the SD indicates the existing dispersion of the results. The AB and SD are indicators of the robustness of the algorithm given to the measurement of the dispersion of the best outputs among themselves, denoting how robust the algorithm is to produce high-quality solutions. The MD indicates the accuracy of the results by ensuring that at least 50% of the solutions obtained by the algorithm or the global optimal solution are the best.

6.5.1 Comparison in Terms of the Quality of Solutions

In this subsection, the performance of the CO-DO is compared against the original CO and seven metaheuristic algorithms employing a set of 28 test functions denoted from $f(x)_1$ to $f(x)_{30}$ which include 14 unimodal functions ($f(x)_1 - f(x)_{14}$), five unimodal functions ($f(x)_{15} - f(x)_{19}$), three hybrid functions ($f(x)_{20} - f(x)_{22}$) and six shifted functions ($f(x)_{23} - f(x)_{28}$), which are detailed in Table 6.9 from the Appendix.

For equal experimentation, all the algorithms consider a population size n of 50 and a dimensionality d of 30, considering equally a stop criterion of 50,000 function accesses for all the algorithms. All the algorithms have been evaluated with a total of 30 independent executions, which means that a total of 30 independent outputs have been obtained for each algorithm per test function. In Table 3.2, the average AB, the median MD and the standard deviation SD for these 30 independent results are summarized and compared.

As can be seen, by the results exposed in Table 6.2, the CO-DO outperforms in 23 of the 30 test functions compared with the other metaheuristic algorithms, also denoting a competitive performance in 4 functions ($f(x)_3, f(x)_4, f(x)_6, f(x)_{12}$) where the global optimum is found for these functions, the same as other metaheuristic algorithms, and only presenting a lower performance in $f(x)_5$ where the SMS presents a better performance. Specifically, the DE presents a competitive mechanism compared to the other. Another interesting discovery in the evaluation is the notorious improvement in the performance with the original CO, where the implementation of the OBL and the diversity-enhanced control potentializes its notoriously competitive performance and significantly enhances the outcomes even in spaces where the optimal solutions are displaced from the origin.

The combination of different strategies selected dynamically based on the evolutionary progress and the enhancement of the explorative capabilities of the OBL affords the algorithm a more uniform and strategic coverage of the space, identifying easily the regions with promising quality and the presence of better preys to focus the hunt. The movement for each dimension following different strategies also generates interesting movement patterns, where the movements are more controlled, generating diverse moving velocities without passing by promising regions but also avoiding the stagnations and slower steps in the early stages of the algorithm, combining both better dynamics and more directed movements. Diversity control also brings more adjusted control and movements, preventing individuals from getting trapped in local optimal regions and preventing the exploration of poor regions at later steps in evolutionary progress. The Comparison in terms of the quality of solutions demonstrates the improved mechanisms with the original CO quality of solutions.

6.5.2 Comparison in Terms of Dimensionality

In this subsection, a comparison of the CO-DO against other metaheuristic approaches is performed, evaluating the optimization performance in a set of 28 test functions evaluated in a more increased number of dimensions d. This analysis is intended to assess the performance of the optimization process of metaheuristic algorithms when they face more challenging landscapes, combining an increase in the number of decision variables to be updated. This represents a greater challenge

Table 6.2 Optimization results for 28 test functions with $d = 30$

$f(x)$		ABC	CS	HS	BA	SMS	PSO	DE	CO	CO-DO
$f(x)_1$	AB	2.28E+00	6.30E+00	1.24E+01	1.54E+01	7.39E+00	4.39E-06	4.52E-08	2.379E-05	**7.22E-15**
	MD	2.26E+00	6.23E+00	1.25E+01	1.58E+01	5.39E+00	3.60E-06	3.45E-08	2.326E-05	**7.11E-15**
	SD	3.10E-01	8.21E-01	5.13E-01	1.22E+00	4.43E+00	3.45E-06	3.70E-08	1.148E-05	**1.47E-15**
$f(x)_2$	AB	5.92E+02	9.87E+00	2.11E+04	7.01E+00	1.54E+01	1.09E+02	6.87E-01	1.201E+00	**6.67E-01**
	MD	5.36E+02	9.28E+00	2.13E+04	6.75E-01	7.51E+00	1.03E+00	6.67E-01	6.668E-01	**6.67E-01**
	SD	3.02E+02	2.26E+00	5.62E+03	1.78E+01	2.42E+01	1.45E+02	7.74E-02	1.224E+00	**3.84E-08**
$f(x)_3$	AB	−3.06E+03	2.34E+09	3.50E+01	−3.72E+06	2.23E+00	1.35E+01	**2.00E+00**	**2.000E+00**	**2.00E+00**
	MD	−1.11E+03	2.00E+00	3.18E+01	−1.56E-01	2.00E+00	9.00E+00	**2.00E+00**	**2.000E+00**	**2.00E+00**
	SD	5.03E+03	4.30E+09	1.48E+01	9.70E+06	1.28E+00	2.04E+01	**0.00E+00**	**0.000E+00**	**0.00E+00**
$f(x)_4$	AB	−8.64E+03	3.67E+09	4.16E+01	−3.98E+06	2.30E+00	1.70E+01	**2.00E+00**	**2.000E+00**	**2.00E+00**
	MD	−1.60E+03	2.00E+00	4.36E+01	−1.56E-01	2.00E+00	9.00E+00	**2.00E+00**	**2.000E+00**	**2.00E+00**
	SD	2.90E+04	4.90E+09	1.71E+01	1.02E+07	1.31E+00	2.41E+01	**0.00E+00**	**0.000E+00**	**0.00E+00**
$f(x)_5$	AB	1.56E-01	7.19E-11	1.25E-03	1.13E-10	**0.00E+00**	6.57E-31	6.03E-15	2.186E-21	7.02E-30
	MD	1.64E-01	0.00E+00	1.26E-03	1.84E-15	**0.00E+00**	0.00E+00	0.00E+00	1.258E-22	0.00E+00
	SD	5.59E-02	3.94E-10	3.10E-04	6.20E-10	**0.00E+00**	2.60E-30	3.06E-14	6.466E-21	1.21E-29
$f(x)_6$	AB	**3.00E+01**	3.01E+01	3.49E+01	5.79E+01	3.01E+01	**3.00E+01**	**3.00E+01**	**3.000E+01**	**3.00E+01**
	MD	**3.00E+01**	3.00E+01	3.50E+01	5.70E+01	3.00E+01	**3.00E+01**	**3.00E+01**	**3.000E+01**	**3.00E+01**
	SD	**0.00E+00**	2.54E-01	1.17E+00	5.35E+00	7.30E-01	**0.00E+00**	**0.00E+00**	**0.000E+00**	**0.00E+00**
$f(x)_7$	AB	6.79E+07	1.00E+10	2.95E+09	1.66E+10	7.85E+10	1.39E-05	3.45E-01	4.225E-04	**3.07E-16**
	MD	5.28E+07	1.00E+10	2.84E+09	1.71E+10	7.72E+10	3.68E-06	2.75E-06	2.269E-04	**4.29E-17**
	SD	4.78E+07	0.00E+00	4.97E+08	1.00E+10	2.20E+10	2.59E-05	1.57E+00	5.687E-04	**8.39E-16**
$f(x)_8$	AB	1.19E+01	1.05E+01	1.44E+01	2.15E+01	1.14E+01	1.09E+01	9.75E+00	1.064E+01	**9.65E+00**

(continued)

Table 6.2 (continued)

$f(x)$		ABC	CS	HS	BA	SMS	PSO	DE	CO	CO-DO
	MD	1.21E+01	1.05E+01	1.46E+01	1.91E+01	1.10E+01	1.09E+01	9.76E+00	1.066E+01	**9.70E+00**
	SD	7.43E-01	4.76E-01	9.38E-01	1.15E+01	1.33E+00	8.04E-01	3.06E-01	6.809E-01	**3.85E-01**
$f(x)_9$	AB	2.39E+02	5.01E+01	1.63E+03	4.98E+01	2.03E+01	1.30E-04	1.60E-04	6.132E-04	**6.43E-08**
	MD	2.03E+02	5.00E+01	1.57E+03	4.97E+01	2.08E+01	1.32E-05	2.01E-05	3.588E-04	**1.57E-10**
	SD	1.21E+02	4.14E+00	4.18E+02	1.90E+01	2.90E+00	3.73E-04	3.79E-04	6.600E-04	**3.17E-07**
$f(x)_{10}$	AB	2.23E+02	1.13E+02	9.59E+01	9.09E+01	6.71E+01	5.73E+01	1.51E+02	1.210E+01	**5.98E-01**
	MD	2.25E+02	1.15E+02	9.64E+01	8.61E+01	5.93E+01	5.57E+01	1.50E+02	1.146E+01	**7.99E-12**
	SD	1.80E+01	1.06E+01	8.46E+00	4.19E+01	2.59E+01	2.09E+01	2.67E+01	3.692E+00	**1.95E+00**
$f(x)_{11}$	AB	1.85E+26	1.00E+10	1.55E+02	1.37E+31	5.13E+33	2.18E+02	7.11E-05	6.435E-05	**3.20E-30**
	MD	1.39E+22	1.00E+10	1.56E+02	3.21E+25	9.55E+30	2.12E+02	1.09E-05	6.061E-05	**4.30E-31**
	SD	1.01E+27	0.00E+00	2.52E+01	6.58E+31	1.89E+34	1.18E+02	1.37E-04	2.819E-05	**6.99E-30**
$f(x)_{12}$	AB	8.83E+00	3.69E+01	4.68E+03	1.60E+04	9.67E+03	**0.00E+00**	2.00E-01	**0.000E+00**	**0.00E+00**
	MD	8.00E+00	3.45E+01	4.72E+03	1.54E+04	6.53E+03	**0.00E+00**	0.00E+00	**0.000E+00**	**0.00E+00**
	SD	2.41E+00	1.08E+01	6.85E+02	4.23E+03	5.90E+03	**0.00E+00**	7.61E-01	**0.000E+00**	**0.00E+00**
$f(x)_{13}$	AB	−7.37E+02	−1.02E+03	−1.10E+03	−9.94E+02	−1.02E+03	−1.07E+03	−1.14E+03	−1.15E+03	**−1.16E+03**
	MD	−7.35E+02	−1.02E+03	−1.09E+03	−9.91E+02	−1.02E+03	−1.08E+03	−1.15E+03	−1.17E+03	**−1.16E+03**
	SD	3.48E+01	1.29E+01	1.26E+01	3.79E+01	2.94E+01	2.62E+01	1.71E+01	3.42E+01	**2.77E+01**
$f(x)_{14}$	AB	−2.64E+14	−2.85E+01	−2.79E+01	−2.57E+14	−2.94E+01	−2.86E+01	−2.46E+01	**−3.00E+01**	**−3.00E+01**
	MD	−2.59E+14	−2.86E+01	−2.80E+01	−2.68E+14	−2.99E+01	−2.81E+01	−2.39E+01	**−3.00E+01**	**−3.00E+01**
	SD	4.48E+13	3.52E-01	3.42E-01	5.19E+13	9.98E-01	1.30E+00	2.72E+00	**1.38E-04**	**4.93E-05**
$f(x)_{15}$	AB	2.67E+01	9.18E+01	2.37E+04	6.09E+04	3.76E+04	3.15E+03	1.37E-13	4.62E-08	**1.21E-36**
	MD	2.55E+01	8.36E+01	2.35E+04	5.62E+04	3.01E+04	2.66E-09	6.46E-14	3.09E-08	**1.66E-38**

(continued)

Table 6.2 (continued)

$f(x)$		ABC	CS	HS	BA	SMS	PSO	DE	CO	CO-DO
	SD	1.04E+01	2.89E+01	3.40E+03	2.75E+04	2.15E+04	4.64E+03	1.94E-13	3.75E-08	**4.01E-36**
$f(x)_{16}$	AB	8.81E+02	2.53E+03	6.62E+05	3.45E+06	3.12E+06	2.27E+05	5.81E-12	2.08E-06	**6.71E-35**
	MD	8.51E+02	2.31E+03	6.48E+05	3.56E+06	2.65E+06	2.10E+05	2.42E-12	1.48E-06	**2.67E-36**
	SD	2.35E+02	9.20E+02	1.26E+05	1.58E+06	1.36E+06	1.67E+05	1.30E-11	1.96E-06	**2.29E-34**
$f(x)_{17}$	AB	1.30E-02	3.81E-02	1.29E+01	1.29E-05	1.76E-02	4.82E-13	1.40E-16	3.52E-11	**1.61E-40**
	MD	1.24E-02	3.64E-02	1.30E+01	1.34E-05	1.68E-02	7.76E-14	5.21E-17	2.29E-11	**1.13E-41**
	SD	4.49E-03	9.88E-03	1.96E+00	1.79E-06	5.65E-03	8.98E-13	2.99E-16	3.50E-11	**3.43E-40**
$f(x)_{18}$	AB	5.71E-01	1.91E+00	5.69E+02	3.04E+00	3.29E+00	6.00E+01	2.43E-14	1.60E-09	**1.31E-38**
	MD	5.60E-01	1.90E+00	5.63E+02	3.72E-04	3.30E+00	1.31E-11	1.43E-15	9.11E-10	**5.34E-40**
	SD	1.83E-01	5.28E-01	8.82E+01	1.50E+01	7.08E-01	1.22E+02	1.14E-13	1.74E-09	**3.81E-38**
$f(x)_{19}$	AB	8.19E-04	3.77E-11	3.45E-05	8.99E-09	5.71E-19	1.34E-25	2.02E-11	6.35E-33	**3.44E-62**
	MD	5.03E-04	2.75E-11	2.90E-05	8.99E-09	1.72E-37	1.74E-28	1.70E-18	3.42E-36	**5.44E-69**
	SD	8.35E-04	3.35E-11	1.59E-05	3.57E-09	3.13E-18	4.90E-25	8.82E-11	2.56E-32	**1.78E-61**
$f(x)_{20}$	AB	2.69E+02	1.17E+03	4.45E+03	2.88E+05	1.08E+06	3.35E+02	4.74E-08	9.926E-06	**1.07E-30**
	MD	2.04E+02	1.02E+03	4.41E+03	2.12E+04	8.30E+04	9.36E-07	4.13E-08	9.684E-06	**1.78E-31**
	SD	2.62E+02	7.20E+02	7.93E+02	1.41E+06	5.28E+06	1.83E+03	2.77E-08	5.409E-06	**2.37E-30**
$f(x)_{21}$	AB	2.18E+02	1.63E+02	3.41E+02	4.94E+02	5.96E+02	6.63E+01	3.71E+01	3.866E+01	**2.90E+01**
	MD	2.21E+02	1.64E+02	3.44E+02	4.86E+02	5.72E+02	6.71E+01	3.68E+01	3.761E+01	**2.90E+01**
	SD	2.11E+01	1.09E+01	2.61E+01	1.15E+02	1.42E+02	1.49E+01	8.54E+00	7.586E+00	**2.59E-10**
$f(x)_{22}$	AB	3.45E+02	3.44E+02	3.41E+02	1.29E+03	2.56E+05	8.51E+01	3.35E+01	3.314E+01	**2.90E+01**
	MD	3.34E+02	3.34E+02	3.45E+02	1.09E+03	2.88E+03	5.61E+01	2.90E+01	2.900E+01	**2.90E+01**
	SD	8.76E+01	4.29E+01	2.85E+01	7.38E+02	9.77E+05	1.03E+02	7.27E+00	7.160E+00	**3.68E-14**

(continued)

Table 6.2 (continued)

$f(x)$		ABC	CS	HS	BA	SMS	PSO	DE	CO	CO-DO
$f(x)_{23}$	AB	2.30E+00	6.72E+00	1.23E+01	1.51E+01	7.41E+00	4.58E-05	3.59E-08	2.753E-05	**7.58E-15**
	MD	2.28E+00	6.65E+00	1.25E+01	1.51E+01	5.35E+00	7.63E-06	2.96E-08	2.193E-05	**7.11E-15**
	SD	2.78E-01	1.25E+00	5.82E-01	1.35E+00	4.62E+00	1.83E-04	3.16E-08	1.905E-05	**2.91E-15**
$f(x)_{24}$	AB	5.83E+02	1.12E+00	4.91E+02	2.12E-02	3.19E+01	1.32E+03	8.36E-04	1.270E-02	**7.58E-04**
	MD	5.49E+02	1.03E+00	5.33E+02	7.60E-03	2.21E+01	4.67E+02	7.39E-04	1.025E-02	**7.58E-04**
	SD	2.69E+02	4.60E-01	1.13E+02	6.88E-02	3.81E+01	1.33E+03	4.58E-04	6.478E-03	**3.53E-04**
$f(x)_{25}$	AB	2.31E+02	1.10E+02	9.54E+01	8.24E+01	6.73E+01	1.33E+02	1.49E+02	1.283E+01	**6.64E-02**
	MD	2.30E+02	1.13E+02	9.63E+01	7.71E+01	6.75E+01	1.36E+02	1.52E+02	1.302E+01	**3.55E-12**
	SD	1.05E+01	1.21E+01	1.03E+01	3.29E+01	2.32E+01	3.10E+01	2.62E+01	3.155E+00	**2.53E-01**
$f(x)_{26}$	AB	2.76E+01	8.32E+01	2.40E+04	6.12E+04	4.12E+04	1.03E+04	2.86E-13	3.408E-08	**0.00E+00**
	MD	2.66E+01	8.36E+01	2.40E+04	5.99E+04	3.15E+04	4.29E+03	6.20E-14	2.456E-08	**0.00E+00**
	SD	9.28E+00	2.14E+01	3.53E+03	3.27E+04	2.86E+04	1.45E+04	6.39E-13	2.964E-08	**0.00E+00**
$f(x)_{27}$	AB	8.56E+02	2.75E+03	6.69E+05	3.99E+06	2.96E+06	1.51E+06	3.60E-12	9.432E-07	**0.00E+00**
	MD	7.26E+02	2.68E+03	6.62E+05	4.10E+06	2.86E+06	1.38E+06	2.39E-12	8.291E-07	**0.00E+00**
	SD	3.17E+02	6.41E+02	9.41E+04	1.68E+06	8.62E+05	6.96E+05	4.07E-12	6.185E-07	**0.00E+00**
$f(x)_{28}$	AB	7.10E+24	1.00E+10	1.50E+02	4.16E+31	1.54E+34	4.77E+02	6.53E-05	4.280E-05	**0.00E+00**
	MD	3.66E+22	1.00E+10	1.44E+02	1.98E+26	4.72E+31	4.00E+02	7.58E-06	3.874E-05	**0.00E+00**
	SD	3.52E+25	0.00E+00	2.63E+01	1.80E+32	6.10E+34	1.68E+02	2.33E-04	2.194E-05	**0.00E+00**

for the optimization algorithms of the increased possibility of stagnations in suboptimal regions. In this section, the CO-DO is compared in a set of 28 benchmarking functions denoted from $f(x)_1$ to $f(x)_{28}$ in 50 and 100 dimensions, considering the population size n of 50 and a stop criterion defined by 50,000 function evaluations. This Comparison is performed against seven metaheuristics from the literature: ABC, CS, HS, BA, SMS, PSO and DE.

6.5.2.1 Comparison in 50 Dimensions

This Comparison employs three important factors that determine both robustness in the quality of the obtained outcomes and accuracy with the expected optimal values: The Average AB, the Median MD and the Standard Deviation SD. For all the test functions, a total of 28 independent executions have been performed per algorithm, obtaining a total of 30 different results per algorithm.

In Table 6.3, the results of the evaluation of 28 functions in 50 dimensions are detailed. As can be seen by the results summarized in Table 6.3, the CO-DO scheme outperforms its competitors in 24 of the employed test functions, presenting a competitive performance in two test functions. ($f(x)_3, f(x)_4$) where the performance equals other algorithms.

For a greater number of dimensions, the dispersion of the individuals and a more objective exploration are fundamental for proper performance on a metaheuristic algorithm by providing diverse information about the aggregation of a certain fraction of the population in suboptimal regions, which represents a challenge in high-dimensional spaces. Our proposed CO-DO combines a diversification technique that efficiently prevents this aggregation of individuals by a strategic separation of the individuals and acquiring new positions along the space, which provides a new spectrum of feasible solutions, a great advantage for spaces where the optimal solution is placed outside from the origin, testing exhaustively the explorative capabilities from the algorithm and preventing possible attractors to the origin. Our proposed hybrid scheme represents a powerful mechanism even for the shifted study cases, denoting an effective exploration of the space and the prevention of the attraction of the individuals to determined regions provided by the original CO and improved by the hybridization of exploration enhancement mechanisms.

6.5.2.2 Comparison in 100 Dimensions

A more rigorous examination of the performance of the hybrid CO-DO is performed by evaluating the optimization performance for diverse problems in 100-dimensional spaces. The evaluation in high-dimensional spaces frequently represents a great difficulty for the optimization given the high quantity of decision variables to be updated, resulting in slower convergence of the individuals and a critical decline in the performance of several metaheuristic algorithms.

Table 6.3 Optimization results for 28 test functions with $d = 50$

$f(x)$		ABC	CS	HS	BA	SMS	PSO	DE	**CO-DO**
$f(x)_1$	AB	1.68E+01	9.11E+00	1.61E+01	1.57E+01	1.91E+01	6.69E-01	2.87E-01	**7.13E-08**
	MD	1.69E+01	8.86E+00	1.61E+01	1.56E+01	1.99E+01	5.19E-01	6.96E-04	**4.09E-08**
	SD	7.15E-01	1.46E+00	3.62E-01	8.92E-01	3.24E+00	7.25E-01	4.84E-01	**6.94E-08**
$f(x)_2$	AB	1.25E+06	2.10E+02	3.60E+05	2.04E+02	5.47E+01	9.38E+03	5.00E+00	**8.96E-01**
	MD	1.12E+06	1.81E+02	3.60E+05	1.15E+00	3.44E+01	9.83E+00	4.91E+00	**6.67E-01**
	SD	3.38E+05	8.60E+01	6.85E+04	6.92E+02	6.70E+01	2.90E+04	2.87E+00	**8.60E-01**
$f(x)_3$	AB	−1.64E+03	1.00E+10	2.58E+06	−7.53E+05	7.56E+04	2.97E+05	**2.00E+00**	**2.00E+00**
	MD	−7.73E+02	1.00E+10	2.51E+06	−1.56E-01	6.25E+02	6.25E+02	**2.00E+00**	**2.00E+00**
	SD	2.45E+03	0.00E+00	1.81E+06	1.23E+06	3.82E+05	7.19E+05	**5.64E-10**	**0.00E+00**
$f(x)_4$	AB	−1.49E+03	1.00E+10	3.02E+06	−1.28E+07	1.41E+03	1.78E+05	**2.00E+00**	**2.00E+00**
	MD	−4.37E+02	1.00E+10	1.84E+06	−1.56E-01	4.09E+02	2.15E+03	**2.00E+00**	**2.00E+00**
	SD	3.89E+03	0.00E+00	3.22E+06	6.27E+07	2.59E+03	5.30E+05	**1.43E-09**	**0.00E+00**
$f(x)_5$	AB	4.62E-01	5.17E-05	1.34E-02	2.39E-04	**0.00E+00**	**0.00E+00**	2.24E-10	6.28E-23
	MD	4.62E-01	0.00E+00	1.31E-02	1.87E-05	**0.00E+00**	**0.00E+00**	4.64E-17	3.71E-24
	SD	7.90E-02	1.07E-04	2.83E-03	3.78E-04	**0.00E+00**	**0.00E+00**	1.21E-09	1.61E-22
$f(x)_6$	AB	4.21E+01	3.49E+01	5.58E+01	7.67E+01	3.11E+01	3.11E+01	3.04E+01	**3.00E+01**
	MD	4.20E+01	3.50E+01	5.60E+01	7.55E+01	3.00E+01	3.00E+01	3.00E+01	**3.00E+01**
	SD	1.73E+00	1.68E+00	2.30E+00	7.73E+00	1.90E+00	2.47E+00	6.15E-01	**0.00E+00**
$f(x)_7$	AB	1.29E+11	1.00E+10	2.75E+10	3.36E+10	2.26E+11	6.73E+02	3.35E+03	**5.98E-03**
	MD	1.31E+11	1.00E+10	2.83E+10	3.11E+10	2.27E+11	4.25E+02	4.11E+01	**2.65E-04**
	SD	2.86E+10	0.00E+00	5.55E+09	1.83E+10	2.78E+10	7.62E+02	1.17E+04	**2.91E-02**
$f(x)_8$	AB	1.14E+02	2.02E+01	4.89E+01	5.18E+01	2.20E+01	2.27E+01	**1.89E+01**	1.95E+01

(continued)

Table 6.3 (continued)

$f(x)$		ABC	CS	HS	BA	SMS	PSO	DE	**CO-DO**
	MD	1.14E+02	2.03E+01	4.93E+01	5.42E+01	2.23E+01	2.23E+01	**1.89E+01**	1.96E+01
	SD	2.24E+01	7.04E-01	3.71E+00	1.49E+01	1.65E+00	2.04E+00	**5.87E-01**	8.27E-01
$f(x)_9$	AB	1.78E+05	1.38E+02	2.24E+04	8.95E+01	4.81E+01	1.37E+00	2.64E-02	**9.59E-04**
	MD	1.73E+05	1.37E+02	2.20E+04	9.24E+01	4.61E+01	3.60E-01	1.75E-02	**3.05E-04**
	SD	4.69E+04	8.04E+00	3.95E+03	2.65E+01	8.32E+00	3.08E+00	3.11E-02	**1.34E-03**
$f(x)_{10}$	AB	5.56E+02	2.38E+02	2.83E+02	1.48E+02	1.56E+02	1.68E+02	3.19E+02	**3.27E+01**
	MD	5.52E+02	2.36E+02	2.89E+02	1.40E+02	1.65E+02	1.64E+02	3.21E+02	**2.85E+01**
	SD	3.38E+01	2.10E+01	1.48E+01	5.41E+01	5.56E+01	3.66E+01	3.55E+01	**1.40E+01**
$f(x)_{11}$	AB	1.05E+58	1.00E+10	2.30E+12	5.09E+56	8.86E+60	7.49E+02	6.69E-02	**4.21E-12**
	MD	8.09E+56	1.00E+10	1.78E+08	1.31E+52	2.79E+58	7.37E+02	1.01E-02	**1.76E-12**
	SD	2.16E+58	0.00E+00	7.14E+12	2.51E+57	3.50E+61	1.37E+02	1.36E-01	**6.87E-12**
$f(x)_{12}$	AB	1.11E+04	3.73E+02	2.30E+04	3.23E+04	2.73E+04	2.47E+00	5.50E+00	**1.10E+00**
	MD	1.11E+04	3.63E+02	2.29E+04	3.19E+04	2.77E+04	2.00E+00	2.00E+00	**1.00E+00**
	SD	2.58E+03	7.92E+01	2.13E+03	9.21E+03	6.96E+03	1.85E+00	1.18E+01	**1.03E+00**
$f(x)_{13}$	AB	−8.05E+02	−1.58E+03	−1.62E+03	−1.65E+03	−1.69E+03	−1.72E+03	**−1.88E+03**	−1.83E+03
	MD	−7.95E+02	−1.58E+03	−1.62E+03	−1.65E+03	−1.67E+03	−1.73E+03	**−1.89E+03**	−1.84E+03
	SD	8.04E+01	1.37E+01	3.12E+01	5.63E+01	4.80E+01	4.34E+01	**2.64E+01**	4.81E+01
$f(x)_{14}$	AB	−3.18E+14	−4.18E+01	−4.14E+01	−3.80E+14	−4.77E+01	−4.34E+01	−3.65E+01	**−5.00E+01**
	MD	−3.08E+14	−4.18E+01	−4.12E+01	−3.93E+14	−4.80E+01	−4.33E+01	−3.61E+01	**−5.00E+01**
	SD	7.11E+13	7.25E-01	1.05E+00	8.37E+13	1.77E+00	3.14E+00	6.61E+00	**1.78E-06**
$f(x)_{15}$	AB	9.01E+04	2.26E+03	1.95E+05	2.63E+05	1.85E+05	2.05E+04	6.16E-05	**1.75E-11**
	MD	8.92E+04	2.29E+03	2.01E+05	2.50E+05	1.83E+05	1.93E+04	9.96E-06	**4.00E-12**

(continued)

Table 6.3 (continued)

$f(x)$		ABC	CS	HS	BA	SMS	PSO	DE	**CO-DO**
	SD	1.49E+04	4.45E+02	2.16E+04	1.00E+05	7.28E+04	1.86E+04	1.58E-04	**6.04E-11**
$f(x)_{16}$	AB	4.95E+06	1.24E+05	1.08E+07	1.77E+07	2.06E+07	2.40E+06	1.60E-03	**3.33E-10**
	MD	4.84E+06	1.27E+05	1.06E+07	1.87E+07	1.93E+07	1.73E+06	7.12E-04	**1.06E-10**
	SD	7.18E+05	2.91E+04	9.61E+05	8.24E+06	6.36E+06	1.56E+06	2.39E-03	**8.09E-10**
$f(x)_{17}$	AB	3.07E+01	6.79E-01	6.12E+01	4.44E-05	5.43E-02	1.03E-05	2.28E-08	**5.70E-15**
	MD	3.05E+01	6.76E-01	6.09E+01	4.37E-05	5.38E-02	5.05E-06	3.12E-09	**1.68E-15**
	SD	4.92E+00	1.18E-01	5.27E+00	4.49E-06	8.57E-03	1.54E-05	6.58E-08	**8.82E-15**
$f(x)_{18}$	AB	2.12E+03	5.53E+01	4.61E+03	1.44E+02	1.68E+01	4.00E+02	1.04E-06	**2.42E-13**
	MD	2.10E+03	5.55E+01	4.66E+03	5.90E+00	1.67E+01	1.50E+02	4.77E-07	**1.59E-13**
	SD	3.31E+02	1.44E+01	5.14E+02	2.57E+02	3.71E+00	5.19E+02	1.83E-06	**2.85E-13**
$f(x)_{19}$	AB	3.63E-01	3.50E-10	2.95E-04	1.10E-08	1.06E-14	2.62E-16	2.55E-08	**1.40E-31**
	MD	3.57E-01	2.03E-10	2.94E-04	9.86E-09	3.27E-20	1.34E-16	1.83E-12	**5.58E-34**
	SD	2.03E-01	5.29E-10	1.16E-04	4.17E-09	5.73E-14	3.65E-16	1.35E-07	**6.75E-31**
$f(x)_{20}$	AB	2.47E+08	8.79E+09	1.98E+04	6.77E+12	2.82E+11	9.04E+03	7.93E-04	**2.55E-11**
	MD	3.77E+06	1.00E+10	1.98E+04	4.06E+04	1.50E+08	1.00E+04	4.52E-04	**1.07E-11**
	SD	5.84E+08	3.16E+09	2.18E+03	3.71E+13	8.99E+11	8.06E+03	9.22E-04	**5.14E-11**
$f(x)_{21}$	AB	1.29E+03	3.52E+02	1.01E+03	9.95E+02	1.53E+03	1.59E+02	1.09E+02	**4.90E+01**
	MD	1.30E+03	3.55E+02	1.00E+03	9.33E+02	1.49E+03	1.56E+02	1.08E+02	**4.90E+01**
	SD	1.15E+02	1.44E+01	6.87E+01	2.29E+02	2.92E+02	2.63E+01	2.54E+01	**2.79E-06**
$f(x)_{22}$	AB	2.35E+08	7.25E+09	1.03E+03	1.26E+07	1.79E+13	5.92E+02	9.44E+01	**4.90E+01**
	MD	8.00E+06	1.00E+10	1.03E+03	1.69E+03	5.81E+07	4.76E+02	8.95E+01	**4.90E+01**
	SD	6.98E+08	4.46E+09	4.80E+01	6.60E+07	6.21E+13	4.27E+02	2.19E+01	**1.98E-05**

(continued)

Table 6.3 (continued)

$f(x)$		ABC	CS	HS	BA	SMS	PSO	DE	CO-DO
$f(x)_{23}$	AB	1.67E+01	9.04E+00	1.60E+01	1.53E+01	1.86E+01	5.33E+00	3.03E-01	**9.79E-08**
	MD	1.68E+01	9.16E+00	1.61E+01	1.53E+01	1.99E+01	1.77E+00	5.87E-04	**7.36E-08**
	SD	7.00E-01	1.25E+00	3.09E-01	8.11E-01	4.14E+00	5.64E+00	5.30E-01	**8.39E-08**
$f(x)_{24}$	AB	1.05E+04	2.39E+01	3.82E+03	1.60E-01	3.92E+01	6.94E+03	6.11E-02	**5.72E-02**
	MD	1.03E+04	2.27E+01	3.78E+03	3.53E-02	1.50E+00	6.71E+03	3.52E-02	**4.92E-02**
	SD	1.98E+03	6.97E+00	5.39E+02	4.90E-01	5.85E+01	2.80E+03	9.28E-02	**3.18E-02**
$f(x)_{25}$	AB	5.59E+02	2.34E+02	2.82E+02	1.38E+02	1.74E+02	3.53E+02	3.23E+02	**3.17E+01**
	MD	5.63E+02	2.33E+02	2.82E+02	1.37E+02	1.78E+02	3.56E+02	3.23E+02	**2.94E+01**
	SD	3.26E+01	2.12E+01	1.56E+01	5.30E+01	3.90E+01	4.51E+01	3.63E+01	**1.42E+01**
$f(x)_{26}$	AB	9.29E+04	2.29E+03	1.97E+05	2.04E+05	1.83E+05	8.23E+04	3.93E-05	**4.27E-10**
	MD	9.17E+04	2.17E+03	1.99E+05	2.03E+05	1.67E+05	8.16E+04	2.20E-05	**1.34E-10**
	SD	1.46E+04	4.49E+02	1.94E+04	1.21E+05	7.71E+04	5.44E+04	3.90E-05	**9.99E-10**
$f(x)_{27}$	AB	5.05E+06	1.21E+05	1.06E+07	1.82E+07	2.21E+07	1.35E+07	3.09E-03	**4.04E-12**
	MD	4.84E+06	1.20E+05	1.06E+07	1.72E+07	2.09E+07	1.35E+07	9.29E-04	**1.53E-12**
	SD	7.90E+05	3.10E+04	1.39E+06	7.19E+06	6.08E+06	5.10E+06	7.40E-03	**5.98E-12**
$f(x)_{28}$	AB	8.78E+57	1.00E+10	2.28E+12	1.74E+60	4.06E+61	9.52E+02	7.70E-02	**4.75E-15**
	MD	5.39E+56	1.00E+10	8.80E+08	5.66E+48	5.17E+57	9.03E+02	7.07E-03	**2.03E-15**
	SD	1.83E+58	0.00E+00	1.11E+13	9.52E+60	2.12E+62	2.30E+02	1.44E-01	**6.66E-15**

In this Comparison, the hybrid scheme CO-DO is compared against seven well-established metaheuristic algorithms from the literature, which are the ABC, CS, HS, BA, SMS, PSO and DE, employing a set of 28 test functions $(f(x)_1 - f(x)_{28})$ Including multimodal, unimodal, hybrid and shifted functions. This Comparison applies the Average Best AB, the Median MD and the Standard Deviation SD for a set of 30 different outcomes produced by each algorithm in all the employed functions.

Table 6.4 enlists the results obtained during the evaluation of the CO-DO with seven different metaheuristic approaches in a set of 28 test functions. As can be seen by the results obtained in Table 6.4, the proposed CO-DO presents a notorious performance in 23 of the employed 28 test functions, representing approximately 82% of the employed test set.

The combination of diverse strategies for all the decision variables proposed by the CO combined with the OBL learning considering new information of the search space and dynamic movements considering different factors based on the position of both current and OBL solutions notoriously facilitates the evolutionary process even for high-dimensional spaces, avoiding the location of suboptimal positions, preventing slow steps and improving the converge rate of the individuals along all the process of the algorithm. The integration of two distinct positions through the OBL improves the algorithm's capability for exploring the search space, even for highly intricate scenarios. Moreover, managing differences reduces the issues presented in high-dimensionality optimization problems, assisting in the prevention of the potential aggregation of individuals in suboptimal regions and poor exploration rates along the process of the algorithm. An interesting behavior is found in the shifted functions, where even with the optimal solution displaced from its original position, the hybrid CO-DO presents a significant capability to locate the region where the global optimal solution is placed and refine the solutions to obtain the best quality solution for most of the proposed shifted test functions employed in the Comparison.

6.5.3 Comparison in the CEC 2017 Benchmark Test Set

An interesting part of the metaheuristic methodologies is the applicability of an optimization mechanism for diverse optimization problems, combining diverse degrees of complexity and the combination of constraints and composition of diverse test functions to generate a competitive set of tests for validating the effectiveness of the optimization process. For many researchers, the CEC 2017 test functions benchmarking set [34] is a set of different test functions frequently employed for diverse competitions and for validating the performance of novel metaheuristic approaches given to the combination of diverse test functions, and the combination of constraints and conditions which represent a challenge for most of the metaheuristic optimizers.

Table 6.4 Optimization results for 28 test functions with $d = 100$

$f(x)$		ABC	CS	HS	BA	SMS	PSO	DE	**CO-DO**
$f(x)_1$	AB	2.07E+01	1.11E+01	1.86E+01	1.55E+01	1.99E+01	5.19E+00	3.01E+00	**9.95E-02**
	MD	2.07E+01	1.09E+01	1.86E+01	1.54E+01	1.99E+01	4.89E+00	2.88E+00	**8.87E-02**
	SD	7.55E-02	1.27E+00	1.57E-01	1.02E+00	7.97E-04	1.97E+00	6.17E-01	**6.15E-02**
$f(x)_2$	AB	2.69E+07	8.04E+03	5.16E+06	5.37E+04	8.25E+02	2.33E+05	2.44E+02	**9.57E+01**
	MD	2.67E+07	7.69E+03	5.34E+06	1.23E+04	7.05E+02	1.77E+05	2.01E+02	**8.31E+01**
	SD	1.81E+06	2.77E+03	5.73E+05	1.22E+05	5.13E+02	2.52E+05	1.56E+02	**4.39E+01**
$f(x)_3$	AB	−1.75E+03	1.00E+10	3.13E+31	−2.43E+07	5.65E+34	1.82E+23	2.00E+00	**2.00E+00**
	MD	−3.42E+02	1.00E+10	2.06E+30	−1.56E-01	2.12E+30	2.92E+16	2.00E+00	**2.00E+00**
	SD	3.43E+03	0.00E+00	8.51E+31	1.29E+08	2.12E+35	9.56E+23	7.61E-05	**0.00E+00**
$f(x)_4$	AB	−1.52E+03	1.00E+10	1.05E+31	−6.47E+05	8.03E+35	5.18E+26	2.00E+00	**2.00E+00**
	MD	−4.21E+02	1.00E+10	9.82E+29	−1.56E-01	1.17E+30	4.29E+18	2.00E+00	**2.00E+00**
	SD	5.07E+03	0.00E+00	2.16E+31	2.03E+06	4.10E+36	2.83E+27	3.99E-04	**0.00E+00**
$f(x)_5$	AB	7.94E-01	6.17E-03	7.85E-02	6.04E-03	**0.00E+00**	**0.00E+00**	8.16E-09	5.06E-07
	MD	7.94E-01	7.84E-03	7.86E-02	3.06E-04	**0.00E+00**	**0.00E+00**	6.70E-10	1.41E-09
	SD	1.09E-01	3.98E-03	9.47E-03	8.59E-03	**0.00E+00**	**0.00E+00**	1.61E-08	2.76E-06
$f(x)_6$	AB	2.14E+02	6.00E+01	1.32E+02	1.28E+02	4.25E+01	4.90E+01	3.75E+01	**3.13E+01**
	MD	2.14E+02	5.95E+01	1.33E+02	1.27E+02	4.15E+01	4.75E+01	3.60E+01	**3.10E+01**
	SD	8.55E+00	3.36E+00	3.81E+00	1.50E+01	4.88E+00	9.58E+00	4.14E+00	**1.12E+00**
$f(x)_7$	AB	9.02E+11	1.00E+10	1.98E+11	6.73E+10	6.20E+11	9.93E+07	1.30E+07	**1.60E+04**
	MD	9.19E+11	1.00E+10	2.02E+11	6.28E+10	6.33E+11	8.28E+07	4.18E+06	**1.64E+04**
	SD	8.14E+10	0.00E+00	1.63E+10	2.68E+10	7.82E+10	6.72E+07	3.04E+07	**4.21E+03**
$f(x)_8$	AB	1.91E+03	**4.62E+01**	3.98E+02	1.93E+02	5.96E+01	7.40E+01	4.77E+01	5.15E+01

(continued)

Table 6.4 (continued)

$f(x)$		ABC	CS	HS	BA	SMS	PSO	DE	**CO-DO**
	MD	1.93E+03	**4.61E+01**	3.99E+02	1.82E+02	6.06E+01	7.09E+01	4.74E+01	5.08E+01
	SD	1.41E+02	**1.30E+00**	2.52E+01	6.62E+01	4.38E+00	1.54E+01	2.38E+00	3.81E+00
$f(x)_9$	AB	1.18E+06	4.34E+02	2.03E+05	4.53E+02	1.68E+02	2.48E+02	2.78E+01	**1.94E+01**
	MD	1.17E+06	4.29E+02	1.98E+05	2.40E+02	1.60E+02	2.32E+02	2.67E+01	**1.61E+01**
	SD	1.08E+05	3.13E+01	1.93E+04	6.60E+02	3.53E+01	7.82E+01	1.44E+01	**9.60E+00**
$f(x)_{10}$	AB	1.60E+03	6.14E+02	8.87E+02	2.36E+02	3.49E+02	4.62E+02	7.51E+02	**1.20E+02**
	MD	1.60E+03	6.14E+02	8.87E+02	2.27E+02	3.43E+02	4.65E+02	7.71E+02	**1.16E+02**
	SD	4.63E+01	3.44E+01	2.06E+01	5.45E+01	5.68E+01	6.09E+01	8.79E+01	**2.84E+01**
$f(x)_{11}$	AB	6.70E+138	1.00E+10	1.37E+76	7.94E+116	3.20E+134	2.19E+03	5.77E+01	**3.28E-02**
	MD	6.59E+135	1.00E+10	1.05E+71	1.18E+107	1.01E+125	2.18E+03	2.16E+01	**2.91E-02**
	SD	2.25E+139	0.00E+00	7.10E+76	3.92E+117	1.68E+135	2.27E+02	1.06E+02	**1.49E-02**
$f(x)_{12}$	AB	2.56E+05	3.44E+03	9.62E+04	6.79E+04	8.73E+04	2.67E+03	2.21E+02	**2.54E+01**
	MD	2.54E+05	3.50E+03	9.60E+04	6.29E+04	8.21E+04	3.51E+02	1.17E+02	**2.25E+01**
	SD	1.33E+04	6.60E+02	4.36E+03	1.88E+04	1.90E+04	4.28E+03	3.03E+02	**1.20E+01**
$f(x)_{13}$	AB	−1.15E+03	−2.76E+03	−2.63E+03	−3.25E+03	−3.32E+03	−3.36E+03	−3.55E+03	**−3.61E+03**
	MD	−1.13E+03	−2.76E+03	−2.62E+03	−3.24E+03	−3.31E+03	−3.36E+03	−3.55E+03	**−3.60E+03**
	SD	1.43E+02	3.79E+01	4.71E+01	6.79E+01	6.92E+01	1.01E+02	6.91E+01	**5.07E+01**
$f(x)_{14}$	AB	−4.52E+14	−6.39E+01	−6.58E+01	−5.90E+14	−9.46E+01	−8.12E+01	−6.18E+01	**−1.00E+02**
	MD	−4.61E+14	−6.39E+01	−6.55E+01	−5.73E+14	−9.53E+01	−8.10E+01	−5.59E+01	**−1.00E+02**
	SD	1.15E+14	2.08E+00	1.71E+00	1.83E+14	3.09E+00	5.62E+00	1.71E+01	**6.92E-02**
$f(x)_{15}$	AB	4.85E+06	4.91E+04	1.81E+06	9.58E+05	8.68E+05	1.69E+05	7.54E+02	**7.59E+00**
	MD	4.87E+06	4.96E+04	1.83E+06	9.23E+05	8.15E+05	1.49E+05	1.44E+02	**5.76E+00**

(continued)

Table 6.4 (continued)

$f(x)$		ABC	CS	HS	BA	SMS	PSO	DE	**CO-DO**
	SD	3.16E+05	7.13E+03	9.37E+04	4.61E+05	2.53E+05	1.04E+05	1.63E+03	**6.06E+00**
$f(x)_{16}$	AB	5.81E+08	5.80E+06	2.20E+08	1.81E+08	2.57E+08	3.72E+07	2.09E+04	**3.70E+02**
	MD	5.90E+08	5.80E+06	2.23E+08	1.78E+08	2.55E+08	3.29E+07	1.35E+04	**3.22E+02**
	SD	4.76E+07	9.47E+05	1.46E+07	6.95E+07	7.45E+07	1.59E+07	2.08E+04	**2.17E+02**
$f(x)_{17}$	AB	6.89E+02	7.35E+00	2.54E+02	**2.71E-04**	6.56E-01	3.85E+00	4.51E-02	1.57E-03
	MD	6.79E+02	7.29E+00	2.56E+02	**2.25E-04**	6.05E-01	3.91E-01	2.47E-02	1.09E-03
	SD	3.35E+01	1.22E+00	1.19E+01	**1.78E-04**	2.95E-01	9.06E+00	6.80E-02	1.42E-03
$f(x)_{18}$	AB	1.15E+05	1.17E+03	4.15E+04	2.94E+03	2.83E+02	4.38E+03	1.16E+01	**1.48E-01**
	MD	1.14E+05	1.17E+03	4.16E+04	2.20E+03	2.08E+02	4.00E+03	3.64E+00	**1.27E-01**
	SD	8.11E+03	1.62E+02	1.92E+03	3.50E+03	1.75E+02	2.54E+03	2.46E+01	**9.69E-02**
$f(x)_{19}$	AB	1.30E+00	1.54E-09	2.35E-03	1.11E-08	**3.70E-11**	3.40E-08	4.12E-01	2.10E-03
	MD	1.40E+00	7.78E-10	2.31E-03	1.10E-08	**1.69E-13**	4.23E-09	1.04E-07	8.30E-13
	SD	3.83E-01	1.80E-09	8.12E-04	5.26E-09	**1.18E-10**	7.44E-08	6.06E-01	1.15E-02
$f(x)_{20}$	AB	1.10E+39	1.00E+10	9.90E+04	9.71E+24	2.15E+35	7.47E+04	4.59E+01	**4.05E-01**
	MD	2.63E+36	1.00E+10	9.98E+04	1.07E+08	1.88E+26	7.10E+04	2.50E+01	**3.60E-01**
	SD	2.79E+39	0.00E+00	5.08E+03	5.31E+25	1.17E+36	3.23E+04	5.28E+01	**2.21E-01**
$f(x)_{21}$	AB	8.10E+03	9.48E+02	3.48E+03	2.13E+03	3.64E+03	6.03E+02	4.31E+02	**2.00E+02**
	MD	8.16E+03	9.51E+02	3.48E+03	2.19E+03	3.55E+03	5.64E+02	4.25E+02	**1.98E+02**
	SD	4.13E+02	3.31E+01	1.38E+02	4.86E+02	4.75E+02	1.38E+02	5.16E+01	**5.57E+01**
$f(x)_{22}$	AB	4.38E+40	1.00E+10	3.60E+03	9.64E+26	1.37E+33	3.94E+03	4.11E+02	**1.01E+02**
	MD	8.87E+35	1.00E+10	3.63E+03	2.37E+08	9.80E+24	4.01E+03	4.04E+02	**9.95E+01**
	SD	2.37E+41	0.00E+00	2.05E+02	5.28E+27	7.04E+33	1.21E+03	6.74E+01	**4.02E+00**

(continued)

Table 6.4 (continued)

$f(x)$		ABC	CS	HS	BA	SMS	PSO	DE	**CO-DO**
$f(x)_{23}$	AB	2.07E+01	1.10E+01	1.86E+01	1.57E+01	1.99E+01	1.66E+01	2.93E+00	**2.01E-01**
	MD	2.07E+01	1.04E+01	1.86E+01	1.57E+01	1.99E+01	1.68E+01	2.69E+00	**1.14E-01**
	SD	8.63E-02	1.21E+00	1.17E-01	9.96E-01	7.07E-04	2.29E+00	9.53E-01	**2.77E-01**
$f(x)_{24}$	AB	8.89E+04	2.47E+02	2.52E+04	**1.11E+01**	3.16E+01	3.07E+04	1.35E+01	7.18E+01
	MD	8.95E+04	2.44E+02	2.61E+04	**9.89E-01**	2.26E+01	3.03E+04	8.82E+00	5.69E+01
	SD	1.27E+04	5.29E+01	2.40E+03	**2.57E+01**	2.28E+01	5.83E+03	1.26E+01	5.33E+01
$f(x)_{25}$	AB	1.60E+03	6.27E+02	8.74E+02	2.57E+02	3.57E+02	8.80E+02	7.42E+02	**1.19E+02**
	MD	1.61E+03	6.27E+02	8.81E+02	2.51E+02	3.63E+02	8.81E+02	7.65E+02	**1.12E+02**
	SD	4.83E+01	3.06E+01	2.50E+01	5.56E+01	5.36E+01	8.10E+01	8.56E+01	**2.41E+01**
$f(x)_{26}$	AB	4.90E+06	5.04E+04	1.74E+06	9.61E+05	7.92E+05	6.43E+05	1.25E+03	**7.14E+00**
	MD	4.87E+06	4.99E+04	1.75E+06	9.85E+05	7.14E+05	6.04E+05	1.34E+02	**4.75E+00**
	SD	3.20E+05	9.12E+03	1.28E+05	3.58E+05	2.77E+05	2.81E+05	4.55E+03	**5.41E+00**
$f(x)_{27}$	AB	5.91E+08	5.56E+06	2.22E+08	1.92E+08	2.80E+08	2.07E+08	3.65E+04	**4.98E+02**
	MD	5.89E+08	5.42E+06	2.24E+08	1.95E+08	2.84E+08	2.00E+08	2.03E+04	**3.37E+02**
	SD	5.78E+07	9.25E+05	1.43E+07	7.62E+07	6.93E+07	5.04E+07	4.63E+04	**5.30E+02**
$f(x)_{28}$	AB	8.86E+138	1.00E+10	3.74E+75	3.76E+134	2.01E+132	3.34E+03	2.63E+01	**4.17E-02**
	MD	1.80E+137	1.00E+10	2.06E+71	6.03E+107	3.77E+127	3.39E+03	1.68E+01	**3.19E-02**
	SD	2.01E+139	0.00E+00	2.02E+76	2.06E+135	9.71E+132	3.81E+02	2.84E+01	**2.57E-02**

For a more extensive validation of the performance, the proposed CO-DO has been tested against the metaheuristic algorithms employed for the validation as the ABC, CS, HS, BA, SMS, PSO and DE employing a representative extract of 14 functions denoted from $f(x)_{29}$ to $f(x)_{42}$ detailed in Table 6.10 from the Appendix. The benchmarking set of the CEC 2017 data set combines diverse characteristics such as the displacement of the global optimal solution, rotation of the search space, and composition with other test functions.

For this validation, a population size n of 50 and a dimensionality d of 100 dimensions are parameters set for all the algorithms. A stop criterion for these test functions has been set on 50,000 function evaluations, and the parameters for each different algorithm have been established as the ones in Table 6.1. Table 6.5 summarizes the statistical results for a total of 30 independent executions, considering the Average AB, the Median MD and the Standard Deviation SD as impact factors of the performance.

The results from Table 6.5 have evidenced a strong exploration capability and the improvement for the refinement of the promising regions even in high-complexity test functions, confirming the reliability of the CO-DO hybrid scheme for a wide spectrum of applications for diverse real-world optimization requirements. The combination of strategies for each decision variable of each individual demonstrates an efficient and dynamical movement for the individuals, taking care of the quality and generating different energy levels for the movement of the individuals, making this optimization scheme easily adaptable for different optimization problems.

6.5.4 *Comparison in Terms of Significance of Differences Between Algorithms*

In this section, a non-parametrical Wilcoxon rank-sum test [35] is performed to validate the significances found in the different behaviors between the original CO-DO with the original CO algorithm and the other metaheuristic approaches employed along the comparison, which are the ABC, CS, HS, BA, SMS, PSO and DE.

In order to perform the Wilcoxon test evaluation, a total of 30 different executions for all the algorithms have been performed along a set of 28 test functions detailed in the Appendix in 30 dimensions. For this evaluation, a significance rate of 5% was considered. For an easier comprehension of the significance results, three different indicators have been employed: ▲, ▶, and ▼. The ▲ symbol indicates an outstanding performance by the CO-DO against another method in terms of the quality of the obtained solutions. The ▶ indicates similarities in the performance between the CO-DO and another method based on the quality of the solutions. The ▲ symbol indicates a minor performance from the CO-DO, given the lower quality of the solutions. In Table 6.5, a summary of the significance values of the CO-DO against the methods employed in the comparison and the improvement indicator is presented.

Table 6.5 Optimization results for 14 CEC2017 test functions with $d = 100$

$f(x)$		ABC	CS	HS	BA	SMS	PSO	DE	**CO-DO**
$f(x)_{29}$	AB	2.75E+06	6.76E+05	6.95E+05	6.86E+06	8.03E+05	8.39E+05	7.20E+05	6.76E+05
	MD	2.57E+06	6.86E+05	7.04E+05	9.26E+05	7.79E+05	8.09E+05	7.22E+05	6.89E+05
	SD	1.06E+06	6.36E+04	5.83E+04	2.75E+07	7.62E+04	1.33E+05	7.66E+04	7.83E+04
$f(x)_{30}$	AB	2.21E+05	3.85E+03	3.74E+04	1.17E+05	3.93E+04	1.44E+04	8.19E+02	**7.69E+02**
	MD	2.21E+05	3.79E+03	3.69E+04	1.16E+05	3.80E+04	1.38E+04	8.10E+02	**7.70E+02**
	SD	2.87E+04	6.94E+02	3.87E+03	3.08E+04	1.27E+04	5.79E+03	5.72E+01	**5.41E+01**
$f(x)_{31}$	AB	2.87E+03	1.73E+03	1.97E+03	2.04E+03	2.33E+03	1.29E+03	1.46E+03	**9.79E+02**
	MD	2.87E+03	1.72E+03	1.97E+03	2.03E+03	2.34E+03	1.29E+03	1.46E+03	**9.75E+02**
	SD	1.16E+02	5.63E+01	2.95E+01	2.03E+02	1.40E+02	8.65E+01	3.86E+01	**7.29E+01**
$f(x)_{32}$	AB	7.56E+02	6.98E+02	6.78E+02	6.89E+02	7.20E+02	6.49E+02	**6.02E+02**	6.15E+02
	MD	7.57E+02	6.98E+02	6.79E+02	6.89E+02	7.19E+02	6.49E+02	**6.02E+02**	6.14E+02
	SD	6.47E+00	4.46E+00	1.97E+00	8.20E+00	6.75E+00	8.15E+00	**8.58E-01**	3.86E+00
$f(x)_{33}$	AB	1.27E+04	2.77E+03	5.71E+03	5.48E+03	6.56E+03	2.44E+03	1.89E+03	**1.66E+03**
	MD	1.27E+04	2.80E+03	5.74E+03	5.43E+03	6.54E+03	2.42E+03	1.89E+03	**1.67E+03**
	SD	7.29E+02	1.28E+02	1.76E+02	7.25E+02	9.44E+02	3.05E+02	5.32E+01	**1.54E+02**
$f(x)_{34}$	AB	3.28E+03	2.06E+03	2.30E+03	2.46E+03	2.71E+03	1.61E+03	1.76E+03	**1.30E+03**
	MD	3.27E+03	2.05E+03	2.30E+03	2.48E+03	2.71E+03	1.59E+03	1.76E+03	**1.28E+03**
	SD	5.47E+01	6.43E+01	3.77E+01	1.42E+02	1.28E+02	1.01E+02	3.31E+01	**7.47E+01**
$f(x)_{35}$	AB	2.13E+05	8.85E+04	7.77E+04	3.02E+04	1.47E+05	6.82E+04	**5.12E+03**	2.13E+04
	MD	2.14E+05	8.78E+04	7.90E+04	3.09E+04	1.50E+05	6.14E+04	**4.83E+03**	1.86E+04
	SD	1.29E+04	1.00E+04	4.84E+03	3.53E+03	1.71E+04	2.46E+04	**2.11E+03**	8.47E+03
$f(x)_{36}$	AB	3.42E+04	8.85E+04	7.77E+04	3.02E+04	1.47E+05	6.82E+04	**5.12E+03**	2.13E+04

(continued)

Table 6.5 (continued)

$f(x)$		ABC	CS	HS	BA	SMS	PSO	DE	CO-DO
	MD	3.43E+04	8.78E+04	7.90E+04	3.09E+04	1.50E+05	6.14E+04	**4.83E+03**	1.86E+04
	SD	4.42E+02	1.00E+04	4.84E+03	3.53E+03	1.71E+04	2.46E+04	**2.11E+03**	8.47E+03
$f(x)_{37}$	AB	1.80E+04	2.45E+04	3.25E+04	2.34E+04	3.14E+04	2.01E+04	3.23E+04	**1.72E+04**
	MD	1.83E+04	2.46E+04	3.27E+04	2.42E+04	3.17E+04	1.96E+04	3.24E+04	**1.71E+04**
	SD	1.01E+03	4.42E+02	6.21E+02	3.10E+03	8.77E+02	3.21E+03	5.59E+02	**1.58E+03**
$f(x)_{38}$	AB	1.01E+05	8.93E+03	1.25E+04	2.20E+04	1.23E+04	8.20E+03	1.02E+04	**5.86E+03**
	MD	9.68E+04	8.98E+03	1.25E+04	2.19E+04	1.21E+04	8.34E+03	1.03E+04	**5.94E+03**
	SD	2.77E+04	3.74E+02	4.55E+02	3.97E+03	1.54E+03	1.02E+03	3.38E+02	**6.40E+02**
$f(x)_{39}$	AB	8.90E+03	6.09E+03	9.25E+03	1.03E+07	3.02E+04	9.53E+03	7.44E+03	**5.44E+03**
	MD	8.94E+03	6.10E+03	9.29E+03	5.54E+06	1.42E+04	8.39E+03	7.47E+03	**5.43E+03**
	SD	3.26E+02	2.82E+02	3.67E+02	1.05E+07	4.01E+04	2.76E+03	2.86E+02	**5.56E+02**
$f(x)_{40}$	AB	5.11E+03	6.11E+03	7.69E+03	6.59E+03	7.49E+03	6.13E+03	7.42E+03	**5.22E+03**
	MD	5.13E+03	6.08E+03	7.70E+03	6.57E+03	7.57E+03	5.96E+03	7.49E+03	**5.18E+03**
	SD	1.33E+02	2.23E+02	2.05E+02	5.04E+02	3.85E+02	6.45E+02	3.95E+02	**4.40E+02**
$f(x)_{41}$	AB	3.60E+04	3.60E+03	3.84E+03	4.86E+03	4.34E+03	3.46E+03	3.29E+03	**2.82E+03**
	MD	3.62E+04	3.62E+03	3.84E+03	4.81E+03	4.34E+03	3.43E+03	3.29E+03	**2.82E+03**
	SD	7.31E+02	7.05E+01	3.77E+01	1.90E+02	1.47E+02	1.70E+02	3.67E+01	**7.84E+01**
$f(x)_{42}$	AB	6.13E+03	2.72E+04	3.43E+04	2.71E+04	3.36E+04	2.38E+04	3.44E+04	**1.91E+04**
	MD	6.18E+03	2.72E+04	3.44E+04	2.77E+04	3.35E+04	2.23E+04	3.45E+04	**1.89E+04**
	SD	2.02E+02	5.22E+02	6.93E+02	3.74E+03	1.19E+03	4.15E+03	5.33E+02	**1.41E+03**

The results obtained from Table 6.6 evidence a significant performance obtained from the CO-DO in terms of performance, even compared with the original CO, denoting a significant improvement from the OBL and diversity management mechanisms for the original algorithm, rejecting the null hypothesis and validating a different and improved performance, demonstrating a superior performance and a reliable alternative for complex optimization problems. The comparison against the original Cheetah Optimizer (CO) has evidenced how the integration of novel strategies has proven to be effective in improving optimization.

6.5.5 Comparison in Terms of Convergence Rapidness and Robustness

In this subsection, the hybrid scheme CO-DO is compared against other metaheuristic algorithms in terms of the individuals' ability to converge in an optimal solution and the robustness of the quality of the solutions. These characteristics for all the algorithms are evaluated by employing two indicators for all the algorithms: the number of function calls (*NFC*) and the acceleration rate (*ARE*).

The number of function calls *NFC* is an indicator of the number of times the algorithm requires to evaluate an individual in the cost function until an expected result is given in the optimization process. An algorithm with a lower *NFC* indicates that the algorithm requires a minor number of evaluations to reach the optimal solution for a given optimization problem. To obtain the *NFC* indicator, a *VTR* (Value-To-Reach) is configured for all the algorithms on different test functions with the main objective of generating equality in the evaluation of the required accesses to the cost function.

In order for the cost functions, the *VTR* proposed for functions $f(x)_2$ was configured in 0.7, for functions $f(x)_{3,4}$ the *VTR* was configured on 2.1, for function $f(x)_8$ the VTR was set to 10, for functions $f(x)_{6,21,22}$ the *VTR* was set to 30, for $f(x)_{24}$ the *VTR* was set to 10^{-3}, for functions $f(x)_{10,25}$ the *VTR* was set to 10^{-1}, the function $f(x)_{13}$ was set to -1100, and for function $f(x)_{14}$ the *VTR* was configured to -30. For the rest of the functions, the *VTR* set value was set to 10^{-6}. For this assessment, an evaluation of the algorithm for a total of 30 different executions in 30 dimensions for the 28 test functions from the traditional benchmarking test was conducted, and the *NFC* numbers given by each of the 30 independent executions were averaged.

Table 6.6 Wilcoxon results for a set of 28 benchmark functions against two different algorithms

	CO-DO versus ABC	CO-DO versus CS	CO-DO versus HS	CO-DO versus BA	CO-DO versus SMS	CO-DO versus PSO	CO-DO versus DE	CO-DO versus CO
$f(x)_1$	2.37E-12▲	2.37E-12▲	2.37E-12▲	2.37E-12▲	2.37E-12▲	2.37E-12▲	2.37E-12▲	2.37E-12▲
$f(x)_2$	3.02E-11▲	3.02E-11▲	3.02E-11▲	3.02E-11▲	3.02E-11▲	4.57E-09▲	3.92E-02▲	3.02E-11▲
$f(x)_3$	1.00E+00▼	6.52E-04▲	1.21E-12▲	3.06E-04▼	8.14E-02▲	1.55E-06▲	1.00E+00▶	1.00E+00▶
$f(x)_4$	1.00E+00▼	6.13E-05▲	1.21E-12▲	1.36E-03▼	4.19E-02▲	2.87E-05▲	1.00E+00▶	1.00E+00▶
$f(x)_5$	1.78E-11▲	5.58E-04▲	1.78E-11▲	1.78E-11▲	6.59E-05▼	6.59E-05▼	6.96E-01▲	1.01E-09▲
$f(x)_6$	1.00E+00▶	1.09E-02▶	9.06E-13▲	1.16E-12▲	1.61E-01▶	1.00E+00▶	1.00E+00▶	1.00E+00▶
$f(x)_7$	3.02E-11▲	1.21E-12▲	3.02E-11▲	3.02E-11▲	3.02E-11▲	3.02E-11▲	3.02E-11▲	3.02E-11▲
$f(x)_8$	3.02E-11▲	1.75E-05▲	3.02E-11▲	6.07E-11▲	1.85E-08▲	2.39E-04▲	1.33E-01▶	4.36E-02▶
$f(x)_9$	3.02E-11▲	3.02E-11▲	3.02E-11▲	3.02E-11▲	3.02E-11▲	3.02E-11▲	3.02E-11▲	3.02E-11▲
$f(x)_{10}$	3.01E-11▲	3.01E-11▲	3.01E-11▲	3.01E-11▲	3.01E-11▲	3.01E-11▲	3.01E-11▲	1.20E-10▲
$f(x)_{11}$	3.02E-11▲	1.21E-12▲	3.02E-11▲	3.02E-11▲	3.02E-11▲	3.02E-11▲	3.02E-11▲	3.02E-11▲
$f(x)_{12}$	1.11E-12▲	1.20E-12▲	1.21E-12▲	1.21E-12▲	1.21E-12▲	3.34E-01▲	1.10E-02▲	3.34E-01▲
$f(x)_{13}$	3.02E-11▲	3.02E-11▲	8.84E-07▶	3.02E-11▲	3.69E-11▲	5.53E-08▲	1.63E-05▶	2.67E-09▶
$f(x)_{14}$	1.55E-11▼	3.02E-11▲	3.02E-11▲	2.85E-04▼	3.02E-11▶	1.95E-03▲	3.02E-11▲	3.02E-11▶
$f(x)_{15}$	3.02E-11▲	3.02E-11▲	3.02E-11▲	3.02E-11▲	3.02E-11▲	3.02E-11▲	3.02E-11▲	3.02E-11▲
$f(x)_{16}$	3.02E-11▲	3.02E-11▲	3.02E-11▲	3.02E-11▲	3.02E-11▲	3.02E-11▲	3.02E-11▲	3.02E-11▲
$f(x)_{17}$	3.02E-11▲	3.02E-11▲	3.02E-11▲	3.02E-11▲	3.02E-11▲	3.02E-11▲	3.02E-11▲	3.02E-11▲
$f(x)_{18}$	3.02E-11▲	3.02E-11▲	3.02E-11▲	3.02E-11▲	3.02E-11▲	3.02E-11▲	3.02E-11▲	3.02E-11▲
$f(x)_{19}$	3.02E-11▲	3.02E-11▲	3.02E-11▲	3.02E-11▲	3.02E-11▲	3.02E-11▲	3.02E-11▲	3.02E-11▲
$f(x)_{20}$	3.02E-11▲	3.02E-11▲	3.02E-11▲	3.02E-11▲	3.02E-11▲	3.02E-11▲	3.02E-11▲	3.02E-11▲
$f(x)_{21}$	3.02E-11▲	3.02E-11▲	3.02E-11▲	3.02E-11▲	3.02E-11▲	3.02E-11▲	1.86E-03▲	3.02E-11▲

(continued)

Table 6.6 (continued)

	CO-DO versus ABC	CO-DO versus CS	CO-DO versus HS	CO-DO versus BA	CO-DO versus SMS	CO-DO versus PSO	CO-DO versus DE	CO-DO versus CO
$f(x)_{22}$	3.02E-11▲	3.02E-11▲	3.02E-11▲	3.02E-11▲	3.02E-11▲	3.02E-11▲	3.02E-11▲	3.02E-11▲
$f(x)_{23}$	6.39E-12▲	6.39E-12▲	6.39E-12▲	6.39E-12▲	6.39E-12▲	6.39E-12▲	6.39E-12▲	6.39E-12▲
$f(x)_{24}$	3.02E-11▲	3.02E-11▲	3.02E-11▲	4.50E-11▲	3.02E-11▲	3.02E-11▲	1.73E-07▼	3.02E-11▲
$f(x)_{25}$	2.52E-11▲	2.52E-11▲	2.52E-11▲	2.52E-11▲	2.52E-11▲	2.52E-11▲	2.52E-11▲	3.58E-10▲
$f(x)_{26}$	1.21E-12▲	1.21E-12▲	1.21E-12▲	1.21E-12▲	1.21E-12▲	1.21E-12▲	1.21E-12▲	1.21E-12▲
$f(x)_{27}$	1.21E-12▲	1.21E-12▲	1.21E-12▲	1.21E-12▲	1.21E-12▲	1.21E-12▲	1.21E-12▲	1.21E-12▲
$f(x)_{28}$	1.21E-12▲	1.69E-14▲	1.21E-12▲	1.21E-12▲	1.21E-12▲	1.21E-12▲	1.21E-12▲	1.21E-12▲
▲	24	27	27	25	25	26	22	22
►	1	1	1	0	2	1	5	6
▼	3	0	0	3	1	1	1	0

Complementary to the analysis of the required number of evaluations in the cost function for algorithms to obtain a desired result, it is important to assess the performance in terms of convergence for two different methodologies. For this purpose, the *ARE* (Acceleration Rate) is a metric that evaluates two different methodologies *A* and *B*, in terms of the convergence of the individuals. For this measurement, the NFC_A and NFC_B both methodologies are compared to determine differences in performance. According to Eq. (6.21), the *ARE* calculation is determined as:

$$ARE_{A/B} = \frac{NFC_A}{NFC_B} \tag{6.21}$$

If the $ARE_{A/B}$ presents a value higher than 1, it indicates that the methodology *B* presents a faster convergence than the method *B* (Tables 6.7 and 6.8).

6.5.6 Analysis of Convergence Graphs

Complementary to the numerical analysis of the behavior and performance of the hybrid approach against the other metaheuristic algorithms, a graphical of the behavior of the algorithms along all the optimization process affords important information about the quality of the solutions and brings a better panorama of the optimization probes from most for all metaheuristics. With this purpose, in this section, an analysis for a representative set of graphs is analyzed. For this analysis, a representative set of 6 functions selected from the set of 28 test functions employed in Sect. 6.5.1 were selected, including four multimodal, a hybrid and a shifted functions and analyzing the improvement of the solutions along 50,000 function evaluations as stop criterion. Since all the algorithms were executed a total of 30 times, the median for all the executions was selected to graphicate the optimization process (Fig. 6.4).

6.6 Conclusions

In this chapter, a novel metaheuristic hybrid approach based on the Cheetah Optimizer (CO) integrating Oppositon-Based Learning (OBL) and diversity management mechanisms, enhancing the exploration of the search space and preventing the stagnation of individuals in local optimal solutions. The Cheetah Optimizer is a nature-inspired metaheuristic algorithm that combines different searching mechanisms inspired by

Table 6.7 Comparison of CO-DO against other metaheuristic algorithms in terms of number of function access

	ABC	CS	HS	BA	SMS	PSO	DE	CO	CO-DO
$f(x)_1$	50,000	50,000	50,000	50,000	50,000	49,958.33	42,061.67	50,000	**18,340.87**
$f(x)_2$	50,000	50,000	50,000	43,590.00	50,000	46,865	24,918.33	42,039.50	**15,441.80**
$f(x)_3$	50,000	37,148.33	50,000	50,000	15,955	50,000	6846.67	7566.83	**3862.53**
$f(x)_4$	50,000	29,258.33	50,000	50,000	13,550	50,000	7121.67	7887	**3862.53**
$f(x)_5$	50,000	22,018.33	50,000	28,430.00	**403.33**	2070	4268.33	20,815.50	8403.80
$f(x)_6$	22,280.00	45,530.00	50,000	50,000	34,863.33	24,685	10,446.67	11,193.50	**4795.27**
$f(x)_7$	50,000.00	50,000	50,000	50,000	50,000	49,928.33	49,315	50,000	**28,204.27**
$f(x)_8$	50,000.00	49,106.67	50,000	50,000	46,575	48,696.67	**33,133.33**	48,760.23	36,310.57
$f(x)_9$	50,000.00	50,000	50,000	50,000	50,000	50,000	50,000	50,000	**33,716.80**
$f(x)_{10}$	50,000	50,000	50,000	50,000	50,000	50,000	50,000	50,000	**39,298.97**
$f(x)_{11}$	50,000	50,000	50,000	50,000	50,000	50,000	49,805	50,000	**15,326.20**
$f(x)_{12}$	50,000	50,000	50,000	50,000	50,000	34,951.67	**13,613.33**	26,537.86	15,159.10
$f(x)_{13}$	50,000	50,000	46,203.33	50,000	50,000	46,853.33	17,071.67	**12,349.50**	**16,086.23**
$f(x)_{14}$	50,000	42,178.33	46,575.00	50,000	23,340	35,865	50,000	50,000	**7135.60**
$f(x)_{15}$	50,000	50,000	50,000	50,000	50,000	46,566.67	30,605	45,411.66	**14,243.87**
$f(x)_{16}$	50,000	50,000	50,000	50,000	50,000	49,891.67	34,988.33	49,314.33	**15,707.00**
$f(x)_{17}$	50,000	50,000	50,000	50,000	50,000	39,313.33	21,928.33	33,387	**10,423.40**
$f(x)_{18}$	50,000	50,000	50,000	50,000	50,000	44,270	26,625	39,628.83	**12,375.00**
$f(x)_{19}$	50,000	25,763.33	50,000	5038.33	18,170	20,658.33	6016.67	10,697.66	**3493.07**
$f(x)_{20}$	50,000	50,000	50,000	50,000	50,000	49,393.33	42,701.67	50,000	**15,513.20**
$f(x)_{21}$	50,000	50,000	50,000	50,000	50,000	50,000	39,290	43,743.03	**7562.87**
$f(x)_{22}$	50,000	50,000	50,000	50,000	50,000	50,000	32,445	35,253.33	**6456.73**

(continued)

Table 6.7 (continued)

	ABC	CS	HS	BA	SMS	PSO	DE	CO	CO-DO
$f(x)_{23}$	50,000	50,000	50,000	50,000	50,000	49,981.67	41,868.33	50,000	**18,609.47**
$f(x)_{24}$	50,000	50,000	50,000	50,000	50,000	50,000	**43,125**	50,000	49,821.93
$f(x)_{25}$	50,000	50,000	50,000	50,000	50,000	50,000	50,000	50,000	**41,294.97**
$f(x)_{26}$	50,000	50,000	50,000	50,000	50,000	49,126.67	30,748.33	45,346.50	**14,232.53**
$f(x)_{27}$	50,000	50,000	50,000	50,000	50,000	50,000	34,521.67	49,897.50	**15,814.67**
$f(x)_{28}$	50,000	50,000	50,000	50,000	50,000	50,000	49,548.33	50,000	**15,169.80**

Table 6.8 Comparison of *ARE* between the CO-DO and other metaheuristic algorithms

	CO-DO versus ABC $ARE_{CO-DO/ABC}$	CO-DO versus CS $ARE_{CO-DO/CS}$	CO-DO versus HS $ARE_{CO-DO/HS}$	CO-DO versus BA $ARE_{CO-DO/BA}$	CO-DO versus SMS $ARE_{CO-DO/SMS}$	CO-DO versus PSO $ARE_{CO-DO/PSO}$	CO-DO versus DE $ARE_{CO-DO/DE}$	CO-DO versus CO $ARE_{CO-DO/CO}$
$f(x)_1$	2.726	2.726	2.726	2.726	2.726	2.724	2.293	2.726
$f(x)_2$	3.238	3.238	3.238	2.823	3.238	3.035	1.614	2.722
$f(x)_3$	12.945	9.618	12.945	12.945	4.131	12.945	1.773	1.959
$f(x)_4$	12.945	7.575	12.945	12.945	3.508	12.945	1.844	2.042
$f(x)_5$	5.950	2.620	5.950	3.383	0.048	0.246	0.508	2.477
$f(x)_6$	4.646	9.495	10.427	10.427	7.270	5.148	2.179	2.334
$f(x)_7$	1.773	1.773	1.773	1.773	1.773	1.770	1.748	1.773
$f(x)_8$	1.377	1.352	1.377	1.377	1.283	1.341	0.912	1.343
$f(x)_9$	1.483	1.483	1.483	1.483	1.483	1.483	1.483	1.483
$f(x)_{10}$	1.272	1.272	1.272	1.272	1.272	1.272	1.272	1.272
$f(x)_{11}$	3.262	3.262	3.262	3.262	3.262	3.262	3.250	3.262
$f(x)_{12}$	3.298	3.298	3.298	3.298	3.298	2.306	0.898	1.751
$f(x)_{13}$	3.108	3.108	2.872	3.108	3.108	2.913	1.061	0.768
$f(x)_{14}$	7.007	5.911	6.527	7.007	3.271	5.026	7.007	7.007
$f(x)_{15}$	3.510	3.510	3.510	3.510	3.510	3.269	2.149	3.188
$f(x)_{16}$	3.183	3.183	3.183	3.183	3.183	3.176	2.228	3.140
$f(x)_{17}$	4.797	4.797	4.797	4.797	4.797	3.772	2.104	3.203
$f(x)_{18}$	4.040	4.040	4.040	4.040	4.040	3.577	2.152	3.202
$f(x)_{19}$	14.314	7.376	14.314	1.442	5.202	5.914	1.722	3.063
$f(x)_{20}$	3.223	3.223	3.223	3.223	3.223	3.184	2.753	3.223

(continued)

Table 6.8 (continued)

	CO-DO versus ABC $ARE_{CO-DO/ABC}$	CO-DO versus CS $ARE_{CO-DO/CS}$	CO-DO versus HS $ARE_{CO-DO/HS}$	CO-DO versus BA $ARE_{CO-DO/BA}$	CO-DO versus SMS $ARE_{CO-DO/SMS}$	CO-DO versus PSO $ARE_{CO-DO/PSO}$	CO-DO versus DE $ARE_{CO-DO/DE}$	CO-DO versus CO $ARE_{CO-DO/CO}$
$f(x)_{21}$	6.611	6.611	6.611	6.611	6.611	6.611	5.195	5.784
$f(x)_{22}$	7.744	7.744	7.744	7.744	7.744	7.744	5.025	5.460
$f(x)_{23}$	2.687	2.687	2.687	2.687	2.687	2.686	2.250	2.687
$f(x)_{24}$	1.004	1.004	1.004	1.004	1.004	1.004	0.866	1.004
$f(x)_{25}$	1.211	1.211	1.211	1.211	1.211	1.211	1.211	1.211
$f(x)_{26}$	3.513	3.513	3.513	3.513	3.513	3.452	2.160	3.186
$f(x)_{27}$	3.162	3.162	3.162	3.162	3.162	3.162	2.183	3.155
$f(x)_{28}$	3.296	3.296	3.296	3.296	3.296	3.296	3.266	3.296

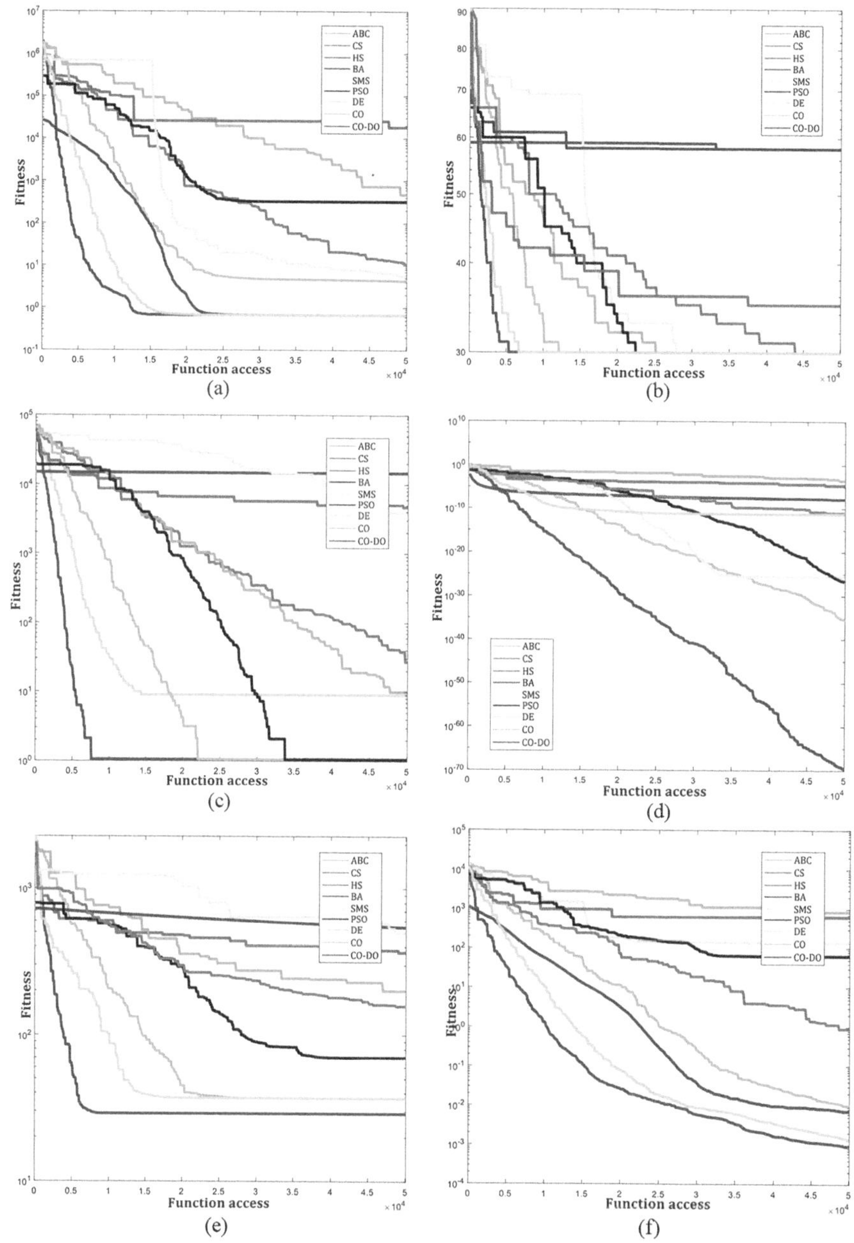

Fig. 6.4 Convergence graphs obtained by the selected algorithms for **a** $f(x)_2$, **b** $f(x)_6$, **c** $f(x)_{12}$, **d** $f(x)_{19}$, **e** $f(x)_{22}$ and **f** $f(x)_{24}$

the hunting behaviors of the cheetahs in their natural environment to capture guardless prey easily and efficiently. The combination of these strategies adapted for the update of the decision variables in the individuals generates interesting movement dynamics with diverse energy levels that prevent both slow movements and poor exploration patterns and uncontrolled movements, preventing a proper refinement of the solutions. Despite the outstanding performance of the original CO for different optimization scenarios, even for real-life requirements, this algorithm presents some difficulties in reaching a proper convergence of the individuals, tending to present some challenges for a full exploration of the search space.

The integration of the OBL for this metaheuristic scheme affords a wider spectrum of information about the solution space by considering two positions simultaneously, presenting different dynamics and movements. On the other hand, the diversity mechanism maintains better control of the convergence of individuals. It maintains a more controlled balance between the exploration and exploitation stages throughout the evolutionary process.

This novel hybrid approach has been experimentally tested in a set of 28 test functions from the traditional benchmarking test set and with a set of 14 functions from the data set of the CEC 2017 test functions against well-established metaheuristic algorithms as the ABC, CS, HS, BA, SMS, PSO, DE and the original CO, evaluating the quality of the solutions, the convergence rapidness and the reliability for all these algorithms. The numerical results have demonstrated a significant improvement against the original CO and an outstanding performance compared to other metaheuristic algorithms, confirming this new hybrid approach for high applicability on diverse optimization problems and application in real-life problems.

Appendix

See Tables 6.9 and 6.10.

Table 6.9 Mathematical representation of employed test functions

	Name	Global minimum	Boundaries	Mathematical function
$f(x)_1$	Ackley	$f(\mathbf{x}^*) = 0$ $\mathbf{x}^* = (0, \ldots, 0)$	$[-30, 30]^d$	$f(x) = -20\exp\left(-0.2\sqrt{\frac{1}{d}\sum_{i=1}^{d} x_i^2}\right) - exp\left(\frac{1}{d}\sum_{i=1}^{d}\cos(2\pi x_i)\right) + 20 + \exp(1)$
$f(x)_2$	Dixon-Price	$f(\mathbf{x}^*) = 0$ $\mathbf{x}^* = 2^{-\frac{2^i-2}{2^i}}$ for $i = 1, \ldots, n$	$[-10, 10]^d$	$f(x) = (x_1 - 1)^2 + \sum_{i=1}^{d} i\left(2x_i^2 - x_{i-1}\right)^2$
$f(x)_3$	Mishra 1	$f(\mathbf{x}^*) = 2$ $\mathbf{x}^* = (1, \ldots, 1)$	$[0, 1]^d$	$f(x) = \left(1 + \left(d - \sum_{i=1}^{d-1} x_i\right)\right)^{d - \sum_{i=1}^{n-1} x_i}$
$f(x)_4$	Mishra 2	$f(\mathbf{x}^*) = 2$ $\mathbf{x}^* = (1, \ldots, 1)$	$[0, 1]^d$	$f(x) = \left(1 + \left(d - \sum_{i=1}^{d-1} \frac{x_i + x_{i+1}}{2}\right)\right)^{d - \sum_{i=1}^{n-1} \frac{x_i + x_{i+1}}{2}}$
$f(x)_5$	Mishra 11	$f(\mathbf{x}^*) = 0$ $\mathbf{x}^* = (0, \ldots, 0)$	$[-10, 10]^d$	$f(x) = \left[\frac{1}{d}\sum_{i=1}^{d} \lvert x_i\rvert - \left(\prod_{i=1}^{d} \lvert x_i\rvert\right)^{\frac{1}{d}}\right]^2$
$f(x)_6$	Plateau	$f(\mathbf{x}^*) = 30$ $\mathbf{x}^* = (0, \ldots, 0)$	$[-5.12, 5.12]^d$	$f(x) = 30 + \sum_{i=1}^{d} \lvert x_i\rvert$
$f(x)_7$	Qing	$f(\mathbf{x}^*) = 0$ $\mathbf{x}^* = (0, \ldots, 0)$	$[-1.28, 1.28]^d$	$f(x) = \sum_{i=1}^{d} \left(x_i^2 - i\right)^2$

(continued)

Table 6.9 (continued)

	Name	Global minimum	Boundaries	Mathematical function
$f(x)_8$	Quartic	$f(\mathbf{x}^*) = 0$ $\mathbf{x}^* = (0, \ldots, 0)$	$[-10, 10]^d$	$f(x) = \sum_{i=1}^{d} ix_i^4 + rand[0, 1]$
$f(x)_9$	Quintic	$f(\mathbf{x}^*) = 0$ $\mathbf{x}^* = (-1, \ldots, -1)$	$[-10, 10]^d$	$f(x) = \sum_{i=1}^{d} \left\| x_i^5 - 3x_i^4 + 4x_i^3 + 2x_i^2 - 10x_i - 4 \right\|$
$f(x)_{10}$	Rastringin	$f(\mathbf{x}^*) = 0$ $\mathbf{x}^* = (1, \ldots, 1)$	$[-5.12, 5.12]^d$	$f(x) = 10d + \sum_{i=1}^{d} \left[x_i^2 - 10\cos(2\pi x_i)\right]$
$f(x)_{11}$	Schweffel 22	$f(\mathbf{x}^*) = 0$ $\mathbf{x}^* = (0, \ldots, 0)$	$[-100, 100]^d$	$f(x) = \sum_{i=1}^{d} \|x_i\| + \prod_{i=1}^{d} \|x_i\|$
$f(x)_{12}$	Step	$f(\mathbf{x}^*) = 0$ $\mathbf{x}^* = (0, \ldots, 0)$	$[-100, 100]^d$	$f(x) = \sum_{i=1}^{d} \left\| x_i^2 \right\|$
$f(x)_{13}$	Styblinski-Tang	$f(\mathbf{x}^*) = -39.1659n$ $\mathbf{x}^* = (-2.90, \ldots, 2.90)$	$[-5, 5]^d$	$f(x) = \frac{1}{2} \sum_{i=1}^{d} \left(x_i^4 - 16x_i^2 + 5x_i\right)$
$f(x)_{14}$	Vincent	$f(\mathbf{x}^*) = -n$ $\mathbf{x}^* = (7.70, \ldots, 7.70)$	$[-0.25, 10]^d$	$f(x) = -\frac{1}{d} \sum_{i=1}^{d} \sin\left[10\log(x_i)\right]$
$f(x)_{15}$	Rothyp	$f(\mathbf{x}^*) = 0$ $\mathbf{x}^* = (0, \ldots, 0)$	$[-65.536, 65.536]^d$	$f(x) = \sum_{i=1}^{d} \sum_{j=1}^{i} x_j^2$
$f(x)_{16}$	Schweffel 2	$f(\mathbf{x}^*) = 0$ $\mathbf{x}^* = (0, \ldots, 0)$	$[-100, 100]^d$	$f(x) = \sum_{i=1}^{d} \left(\sum_{j=1}^{i} x_i\right)^2$

(continued)

Table 6.9 (continued)

	Name	Global minimum	Boundaries	Mathematical function
$f(x)_{17}$	Sphere	$f(\mathbf{x}^*) = 0$ $\mathbf{x}^* = (0, \ldots, 0)$	$[-5, 5]^d$	$f(x) = \sum_{i=1}^{d} x_i^2$
$f(x)_{18}$	Sum of squares	$f(\mathbf{x}^*) = 0$ $\mathbf{x}^* = (0, \ldots, 0)$	$[-10, 10]^d$	$f(x) = \sum_{i=1}^{d} ix_i^2$
$f(x)_{19}$	Sum of different powers	$f(\mathbf{x}^*) = 0$ $\mathbf{x}^* = (0, \ldots, 0)$	$[-1, 1]^d$	$f(x) = \sum_{i=1}^{d} \lvert x_i \rvert^{i+1}$
$f(x)_{20}$	Rastringin+Schweffel 2+Sphere	$f(\mathbf{x}^*) = 0$ $\mathbf{x}^* = (0, \ldots, 0)$	$[-100, 100]^d$	$f(x) = 10n + \sum_{i=1}^{d}[x_i^2 - 10\cos(2\pi x_i)] + \sum_{i=1}^{d}\left(\sum_{j=1}^{i} x_i\right)^2 + \sum_{i=1}^{d} x_i^2$
$f(x)_{21}$	Griewank+Rastringin+Rosenbrock	$f(\mathbf{x}^*) = n - 1$ $\mathbf{x}^* = (0, \ldots, 0)$	$[-100, 100]^d$	$f(x) = \sum_{i=1}^{n} \frac{x_i^2}{4000} - \prod_{i=1}^{n} \cos\left(\frac{x_i}{\sqrt{i}}\right) + 1 + 10n + \sum_{i=1}^{n}[x_i^2 - 10\cos(2\pi x_i)] + \sum_{i=1}^{n-1}[100\left(x_{i+1} - x_i^2\right)^2 + (x_i - 1)^2]$

(continued)

Table 6.9 (continued)

	Name	Global minimum	Boundaries	Mathematical function
$f(x)_{22}$	Ackley+Griewank+Rastringin+Rosenbrock+Schweffel 22	$f(\mathbf{x}^*) = n - 1$ $\mathbf{x}^* = (0, \ldots, 0)$	$[-100, 100]^d$	$f(x) = -20\exp\left(-0.2\sqrt{\frac{1}{n}\sum_{i=1}^{n} x_i^2}\right) - exp\left(\frac{1}{n}\sum_{i=1}^{n}\cos(2\pi x_i)\right) + 20 + \exp(1) + (0.1\{\sin(3\pi x_i) + \sum_{i=1}^{n}(x_i - 1)^2\left[1 + \sin^2(3\pi x_i + 1)\right] + [(x_n - 1)^2[1 + \sin^2(2\pi x_n)]]\} + \sum_{i=1}^{n} u(x_{i,}, 5.100{,}4) + \sum_{i=1}^{n}\left(\sum_{j=1}^{i} x_i\right)^2$
$f(x)_{23}$	Shifted Ackley	$f(\mathbf{x}^*) = 0$ $\mathbf{x}^* = (10, \ldots, 10)$	$[-30, 30]^d$	$f(x) = -20\exp\left(-0.2\sqrt{\frac{1}{d}\sum_{i=1}^{d} z_i^2}\right) - exp\left(\frac{1}{d}\sum_{i=1}^{d}\cos(2\pi z_i)\right) + 20 + \exp(1)$ $z_i = x_i - o_i, \mathbf{o} = [o_1, \ldots o_d], o_i = 10$

(continued)

Table 6.9 (continued)

Name	Global minimum	Boundaries	Mathematical function
Shifted Powell	$f(\mathbf{x}^*) = 0$ $\mathbf{x}^* = (5, \dots, 5)$	$[-4, 5]^d$	$f(x) = \sum_{i=1}^{\frac{d}{4}} [(x_{4i-3} + 10z_{4i-2})^2 + 5(z_{i-1} + z_{4i})^2 + (z_{4i-2} + 2z_{4i-1})^4 + 10(z_{4i-3} + z_{4i})^4]$ $z_i = x_i - o_i, \mathbf{o} = [o_1, \dots o_d], o_i = 5$
Shifted Rastringin	$f(\mathbf{x}^*) = 0$ $\mathbf{x}^* = (15, \dots, 15)$	$[-5.12, 5.12]^d$	$f(x) = 10d + \sum_{i=1}^{d} [z_i^2 - 10\cos(2\pi z_i)]$ $z_i = x_i - o_i, \mathbf{o} = [o_1, \dots o_d], o_i = 15$
Shifted Rothyp	$f(\mathbf{x}^*) = 0$ $\mathbf{x}^* = (20, \dots, 20)$	$[-65.536, 65.536]^d$	$f(x) = \sum_{i=1}^{d} \sum_{j=1}^{i} z_j^2$ $z_i = x_i - o_i, \mathbf{o} = [o_1, \dots o_d], o_i = 15$
Shifted Schweffel 2	$f(\mathbf{x}^*) = 0$ $\mathbf{x}^* =$ $(100, \dots, 100)$	$[-100, 100]^d$	$f(x) = \sum_{i=1}^{d} \left(\sum_{j=1}^{i} z_i \right)^2$ $z_i = x_i - o_i, \mathbf{o} = [o_1, \dots o_d], o_i = 100$
Shifted Schwefel 22	$f(\mathbf{x}^*) = 0$ $\mathbf{x}^* = (25, \dots, 25)$	$[-100, 100]^d$	$f(x) = \sum_{i=1}^{d} \lvert z_i \rvert + \prod_{i=1}^{d} \lvert z_i \rvert$ $z_i = x_i - o_i, \mathbf{o} = [o_1, \dots o_d], o_i = 25$

Table 6.10 CEC-2017 benchmark test functions

	Global minimum	Number	Function
$f(x)_{29}$	$f(\mathbf{x}^*) = 300$	f_3	Shifted and rotated Zakharov function
$f(x)_{30}$	$f(\mathbf{x}^*) = 400$	f_4	Shifted and rotated Rosenbrock's function
$f(x)_{31}$	$f(\mathbf{x}^*) = 500$	f_5	Shifted and rotated Rastringin's function
$f(x)_{32}$	$f(\mathbf{x}^*) = 600$	f_6	Shifted and rotated expanded Scaffer's F6 function
$f(x)_{33}$	$f(\mathbf{x}^*) = 700$	f_7	Shifted and rotated Lunacek bi_ Rastringin's function
$f(x)_{34}$	$f(\mathbf{x}^*) = 800$	f_8	Shifted and rotated non-continuous bi_Rastringin's function
$f(x)_{35}$	$f(\mathbf{x}^*) = 900$	f_9	Shifted and rotated Levy function
$f(x)_{36}$	$f(\mathbf{x}^*) = 1000$	f_{10}	Shifted and rotated Schweffel's function
$f(x)_{37}$	$f(\mathbf{x}^*) = 1600$	f_{16}	Hybrid function 6 (N = 4) expanded Schaffer's F6, hgbat, Rosenbrock's, modified Schweffel's
$f(x)_{38}$	$f(\mathbf{x}^*) = 1700$	f_{17}	Hybrid function 6 (N = 5) Katsuura, Ackley, expanded Griewank's plus, Rosenbrock's, Schweffel's. Rastringin's
$f(x)_{39}$	$f(\mathbf{x}^*) = 2000$	f_{20}	Hybrid function 6 (N = 5) Happycat, Katsuura, Ackley, Rastringin's. modified Schweffel's, Schaffer F7
$f(x)_{40}$	$f(\mathbf{x}^*) = 2100$	f_{21}	Composition function 1 (N = 3) Rosenbrock's, high-conditioned elliptic, Rastringin's
$f(x)_{41}$	$f(\mathbf{x}^*) = 2200$	f_{22}	Composition function 1 (N = 3) Rastringin's, Griewank's, modified Schweffel's
$f(x)_{42}$	$f(\mathbf{x}^*) = 2300$	f_{23}	Composition function 1 (N = 4) Rosenbrock's, Ackley, modified Schweffel's, Rastringin's

References

1. X. Yang, *Engineering Optimization* (Wiley, 2010). https://doi.org/10.1002/9780470640425
2. S. Carbas, A. Toktas, D. Ustun, *Nature-Inspired Metaheuristic Algorithms for Engineering Optimization Applications*. (Springer Singapore, Singapore, 2021). https://doi.org/10.1007/978-981-33-6773-9

3. S. Kaul, Y. Kumar, Nature-inspired metaheuristic algorithms for constraint handling: challenges, issues, and research perspective, in *Constraint Handling in Metaheuristics and Applications* (Springer Singapore, Singapore, 2021), pp. 55–80. https://doi.org/10.1007/978-981-33-6710-4_3
4. S.S. Vinod Chandra, H.S. Anand, Nature inspired meta heuristic algorithms for optimization problems. Computing **104**(2), 251–269 (2022). https://doi.org/10.1007/s00607-021-00955-5
5. X.-S. Yang, *Nature Inspired Metaheuristic Algorithms*, 2nd edn. (Lunvier Press, United Kingdom, 2010)
6. R. Storn, K. Price, Differential evolution – a simple and efficient heuristic for global optimization over continuous spaces. J. Global Optim. **11**(4), 341–359 (1997). https://doi.org/10.1023/A:1008202821328
7. J.H. Holland, Genetic algorithms and adaptation, in *Adaptive Control of Ill-Defined Systems* (Springer US, Boston, MA, 1984), pp. 317–333. https://doi.org/10.1007/978-1-4684-8941-5_21
8. H.-G. Beyer, H.-P. Schwefel, Evolution strategies – a comprehensive introduction. Nat. Comput. **1**(1), 3–52 (2002). https://doi.org/10.1023/A:1015059928466
9. J. Kennedy, R. Eberhart, Particle swarm optimization, in *Proceedings of ICNN'95 - International Conference on Neural Networks*, IEEE, pp. 1942–1948. https://doi.org/10.1109/ICNN.1995.488968
10. S. Mirjalili, A. Lewis, The whale optimization algorithm. Adv. Eng. Softw. **95**, 51–67 (2016). https://doi.org/10.1016/j.advengsoft.2016.01.008
11. S. Mirjalili, S.M. Mirjalili, A. Lewis, Grey wolf optimizer. Adv. Eng. Softw. **69**, 46–61 (2014). https://doi.org/10.1016/j.advengsoft.2013.12.007
12. D. Karaboga, An idea based on honey bee swarm for numerical optimization. Technical report-TR06, Erciyes University, Engineering Faculty (2005)
13. X.-S. Yang, A new metaheuristic bat-inspired algorithm (2010), pp. 65–74. https://doi.org/10.1007/978-3-642-12538-6_6
14. X.-S. Yang, S. Deb, Cuckoo search via lévy flights, in *2009 World Congress on Nature & Biologically Inspired Computing (NaBIC)*, IEEE (2009), pp. 210–214. https://doi.org/10.1109/NABIC.2009.5393690
15. A. Askarzadeh, A novel metaheuristic method for solving constrained engineering optimization problems: crow search algorithm. Comput. Struct. **169**, 1–12 (2016). https://doi.org/10.1016/j.compstruc.2016.03.001
16. S. Kirkpatrick, C.D. Gelatt, M.P. Vecchi, Optimization by simulated annealing. Science (1979) **220**(4598), 671–680 (1983). https://doi.org/10.1126/science.220.4598.671
17. E. Rashedi, H. Nezamabadi-pour, S. Saryazdi, GSA: a gravitational search algorithm. Inf. Sci. (N.Y.) **179**(13), 2232–2248 (2009). https://doi.org/10.1016/j.ins.2009.03.004
18. Z.W. Geem, J.H. Kim, G.V. Loganathan, A new heuristic optimization algorithm: harmony search. SIMULATION **76**(2), 60–68 (2001). https://doi.org/10.1177/003754970107600201
19. E. Cuevas, A. Echavarría, M.A. Ramírez-Ortegón, An optimization algorithm inspired by the states of matter that improves the balance between exploration and exploitation. Appl. Intell. **40**(2), 256–272 (2014). https://doi.org/10.1007/s10489-013-0458-0
20. M. Abdel-Basset, R. Mohamed, S.A.A. Azeem, M. Jameel, M. Abouhawwash, Kepler optimization algorithm: a new metaheuristic algorithm inspired by Kepler's laws of planetary motion. Knowl. Based Syst. **268**, 110454 (2023). https://doi.org/10.1016/j.knosys.2023.110454
21. M.A. Akbari, M. Zare, R. Azizipanah-abarghooee, S. Mirjalili, M. Deriche, The cheetah optimizer: a nature-inspired metaheuristic algorithm for large-scale optimization problems. Sci. Rep. **12**(1), 10953 (2022). https://doi.org/10.1038/s41598-022-14338-z
22. Z.A. Memon, M.A. Akbari, M. Zare, An improved cheetah optimizer for accurate and reliable estimation of unknown parameters in photovoltaic cell and module models. Appl. Sci. **13**(18), 9997 (2023). https://doi.org/10.3390/app13189997
23. Y. Fu, S. Yang, B. Liu, E. Xia, D. Huang, Multi-UAV cooperative trajectory planning based on the modified cheetah optimization algorithm. Entropy **25**(9), 1277 (2023). https://doi.org/10.3390/e25091277

24. T.O. Ting, X.-S. Yang, S. Cheng, K. Huang, Hybrid metaheuristic algorithms: past, present, and future (2015), pp. 71–83. https://doi.org/10.1007/978-3-319-13826-8_4
25. G.R. Raidl, A unified view on hybrid metaheuristics (2006), pp. 1–12. https://doi.org/10.1007/11890584_1
26. S. Chen, A. Bolufe-Rohler, J. Montgomery, T. Hendtlass, An analysis on the effect of selection on exploration in particle swarm optimization and differential evolution, in *2019 IEEE Congress on Evolutionary Computation (CEC)*, IEEE, June 2019 (2019), pp. 3037–3044. https://doi.org/10.1109/CEC.2019.8790200
27. M. Črepinšek, S.-H. Liu, M. Mernik, Exploration and exploitation in evolutionary algorithms. ACM Comput. Surv. **45**(3), 1–33 (2013). https://doi.org/10.1145/2480741.2480752
28. B. Morales-Castañeda, D. Zaldívar, E. Cuevas, F. Fausto, A. Rodríguez, A better balance in metaheuristic algorithms: does it exist? Swarm Evol. Comput. **54**, 100671 (2020). https://doi.org/10.1016/j.swevo.2020.100671
29. Q. Xu, L. Wang, N. Wang, X. Hei, L. Zhao, A review of opposition-based learning from 2005 to 2012. Eng. Appl. Artif. Intell. **29**, 1–12 (2014). https://doi.org/10.1016/j.engappai.2013.12.004
30. H.R. Tizhoosh, Opposition-based learning: a new scheme for machine intelligence, in *International Conference on Computational Intelligence for Modelling, Control and Automation and International Conference on Intelligent Agents, Web Technologies and Internet Commerce (CIMCA-IAWTIC'06)*, IEEE, pp. 695–701. https://doi.org/10.1109/CIMCA.2005.1631345
31. X.-S. Yang, S. Deb, S. Fong, Metaheuristic algorithms: optimal balance of intensification and diversification. Appl. Math. & Inf. Sci. **8**(3), 977–983 (2014). https://doi.org/10.12785/amis/080306
32. J. Xu, J. Zhang, Exploration-exploitation tradeoffs in metaheuristics: survey and analysis, in *Proceedings of the 33rd Chinese Control Conference*, IEEE, July 2014 (2014), pp. 8633–8638. https://doi.org/10.1109/ChiCC.2014.6896450
33. K. Hussain, M.N. Mohd Salleh, S. Cheng, R. Naseem, Common benchmark functions for metaheuristic evaluation: a review. JOIV : Int. J. Inform.Vis. **1**(4–2), 218–223 (2017).https://doi.org/10.30630/joiv.1.4-2.65
34. G. Wu, R. Mallipeddi, P.N. Suganthan, Problem definitions and evaluation criteria for the CEC 2017 competition and special session on constrained single objective real-parameter optimization. Technical report, September 2017 (2017)
35. F. Wilcoxon, Probability tables for individual comparisons by ranking methods. Biometrics **3**(3), 119 (1947). https://doi.org/10.2307/3001946

The manufacturer's authorised representative in the EU is Springer Nature Customer Service Centre GmbH, Europaplatz 3, 69115 Heidelberg, Germany. If you have any concerns regarding our products, please contact ProductSafety@springernature.com

Printed and bound by CPI Group (UK) Ltd, Croydon, CR0 4YY
01/10/2026
02225925-0010